STUDENT SOLUTIONS MANUAL

to accompany

MULTIVARIABLE CALCULUS

William G. McCallum
University of Arizona

Deborah Hughes-Hallett
Harvard University
et al.

Andrew M. Gleason
Harvard University

STONY BOOKS
USED BOOKS

John Wiley & Sons, Inc.

New York   Chichester   Weinheim   Brisbane   Singapore   Toronto

This project was supported, in part,
by the

National Science Foundation

Opinions expressed are those of the authors
and not necessarily those of the Foundation

Grant No. DUE-9352905

ISBN 0-471-17356-8

Printed in the United States of America

10 9 8 7 6 5 4 3 2

Printed and bound by Bradford & Bigelow, Inc.

CONTENTS

CHAPTER ELEVEN

Solutions for Section 11.1

1. (a) 80-90°F (b) 60-72°F (c) 60-100°F

5. (a) Beef consumption by households making $20,000/year is given by Row 1 of Table 11.1

TABLE 11.1

p	3.00	3.50	4.00	4.50
$f(20, p)$	2.65	2.59	2.51	2.43

For households making $20,000/year, beef consumption decreases as price goes up.

(b) Beef consumption by households making $100,000/year is given by Row 5 of Table 11.1

TABLE 11.2

p	3.00	3.50	4.00	4.50
$f(100, p)$	5.79	5.77	5.60	5.53

For households making $100,000/year, beef consumption also decreases as price goes up.

(c) Beef consumption by households when the price of beef is $3.00/lb is given by Column 1 of Table 11.1

TABLE 11.3

I	20	40	60	80	100
$f(I, 3.00)$	2.65	4.14	5.11	5.35	5.79

When the price of beef is $3.00/lb, beef consumption increases as income increases.

(d) Beef consumption by households when the price of beef is $4.00/lb is given by Column 3 of Table 11.1

TABLE 11.4

I	20	40	60	80	100
$f(I, 4.00)$	2.51	3.94	4.97	5.19	5.60

When the price of beef is $4.00/lb, beef consumption increases as income increases.

9. We have $M = f(B, t) = B(1.05)^t$.

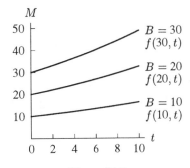

Figure 11.1

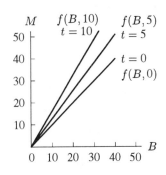

Figure 11.2

Figure 11.1 gives the graphs of f as a function of t for B fixed at 10, 20, and 30. For each fixed B, the function $f(B, t)$ is an increasing function of t. The larger the fixed value of B, the larger $f(B, t)$ is.

Figure 11.2 gives the graphs of f as a function of B for t fixed at 0, 5, and 10. For each fixed t, $f(B, t)$ is an increasing (and in fact linear) function of B. The larger t is, the larger the slope of the line.

13.

TABLE 11.5   *Temperature adjusted for wind-chill at* $20°F$

°F\mph	5	10	15	20	25
20 ° F	16	3	−5	−10	−15

TABLE 11.6   *Temperature adjusted for wind-chill at* $0°F$

°F\mph	5	10	15	20	25
0 ° F	−5	−22	−31	−39	−44

17. (a) For $t = 0$, we have $y = f(x, t) = \sin x, 0 \leq x \leq \pi$

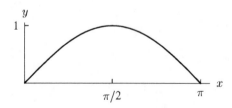

Figure 11.3

For $t = \pi/4$, we have $y = f(x,t) = \frac{\sqrt{2}}{2}\sin x, 0 \le x \le \pi$

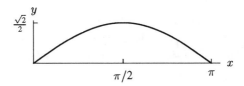

Figure 11.4

For $t = \pi/2$, we have $y = f(x,t) = 0$

Figure 11.5

For $t = 3\pi/4$, we have $y = f(x,t) = \frac{-\sqrt{2}}{2}\sin x, 0 \le x \le \pi$

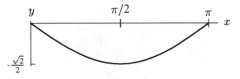

Figure 11.6

For $t = \pi$, we have $y = f(x,t) = -\sin x, 0 \le x \le \pi$

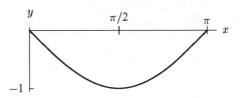

Figure 11.7

(b)   The graphs show an arch of a sine wave which is above the x-axis, concave down at $t = 0$, is straight along the x-axis at $t = \pi/2$, and below the x-axis, concave up at $t = \pi$, like a guitar string vibrating up and down.

Solutions for Section 11.2

1. The distance of a point $P = (x, y, z)$ from the yz-plane is $|x|$, from the xz-plane is $|y|$, and from the xy-plane is $|z|$. So, B is closest to the yz-plane, since it has the smallest x-coordinate in absolute value. B lies on the xz-plane, since its y-coordinate is 0. B is farthest from the xy-plane, since it has the largest z-coordinate in absolute value.

5. The graph is a plane parallel to the yz-plane, and passing through the point $(-3, 0, 0)$. See Figure 11.8.

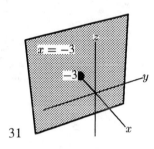

Figure 11.8

9. The equation for the points whose distance from the x-axis is 2 is given by $\sqrt{y^2 + z^2} = 2$, i.e. $y^2 + z^2 = 4$. It specifies a cylinder of radius 2 along the x-axis. See Figure 11.9.

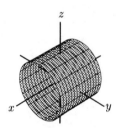

Figure 11.9

13. By drawing the top four corners, we find that the length of the edge of the cube is 5. See Figure 11.10. We also notice that the edges of the cube are parallel to the coordinate axis. So the x-coordinate of the the center equals

$$-1 + \frac{5}{2} = 1.5.$$

The y-coordinate of the center equals

$$-2 + \frac{5}{2} = 0.5.$$

The z-coordinate of the center equals

$$2 - \frac{5}{2} = -0.5.$$

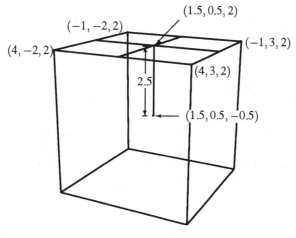

Figure 11.10

17.   An example is the line $y = z$ in the yz-plane. See Figure 11.11.

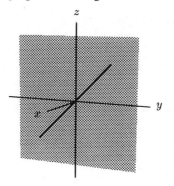

Figure 11.11

Solutions for Section 11.3

1.   (a)   The value of z decreases as x increases. See Figure 11.12.
     (b)   The value of z increases as y increases. See Figure 11.13.

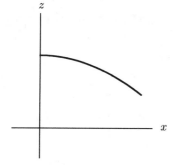

Figure 11.12

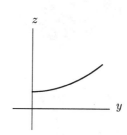

Figure 11.13

5. (a) The value of z only depends on the distance from the point (x, y) to the origin. Therefore the graph has a circular symmetry around the z-axis. There are two such graphs among those depicted in Figure 5: I and V. The one corresponding to $z = \frac{1}{x^2+y^2}$ is I since the function blows up as (x, y) gets close to $(0, 0)$.

    (b) For similar reasons as in part (a), the graph is circularly symmetric about the z-axis, hence the corresponding one must be V.

    (c) The graph has to be a plane, hence IV.

    (d) The function is independent of x, hence the corresponding graph can only be II. Notice that the cross-sections of this graph parallel to the yz-plane are parabolas, which is a confirmation of the result.

    (e) The graph of this function is depicted in III. The picture shows the cross-sections parallel to the zx-plane, which have the shape of the cubic curves $z = x^3 - $ constant.

9. (a)

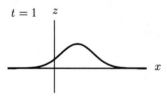

Figure 11.14   *Figure 11.15*

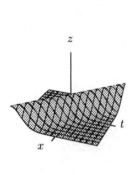

Figure 11.16   *Figure 11.17*

    (b) Increasing x

    (c) The graph in Figure 11.18 represents a wave traveling in the opposite direction.

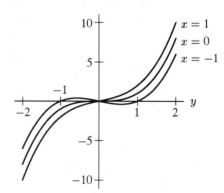

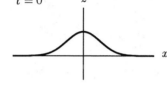

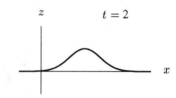

Figure 11.18

Figure 11.19: Cross-section
$f(x, b) = b^3 + bx$, with $b = -1, 0, 1$

13. (a) Cross-sections with x fixed at $x = b$ are in Figure 11.19.

     (b) Cross-section with y fixed at $y = 6$ are in Figure 11.20.

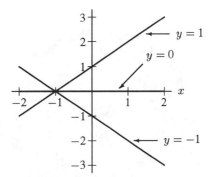

Figure 11.20: Cross-section
$f(x, b) = b^3 + bx$, with $b = -1, 0, 1$

Solutions for Section 11.4

1. The contour where $f(x, y) = x + y = c$, or $y = -x + c$, is the graph of the straight line with slope -1 as shown in Figure 11.21. Note that we have plotted the contours for $c = -3, -2, -1, 0, 1, 2, 3$. The contours are evenly spaced.

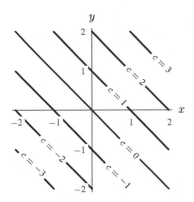

Figure 11.21

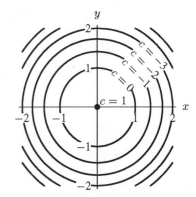

Figure 11.22

5. The contour where $f(x, y) = -x^2 - y^2 + 1 = c$, where $c \leq 1$, is the graph of the circle centered at $(0, 0)$, with radius $\sqrt{1 - c}$ as shown in Figure 11.22. Note that we have plotted the contours for $c = -3, -2, -1, 0, 1$. The contours become more closely packed as we move further from the origin.

9. The contour where $f(x, y) = \cos(\sqrt{x^2 + y^2}) = c$, where $-1 \leq c \leq 1$, is a set of circles centered at $(0, 0)$, with radius $\cos^{-1} c + 2k\pi$ with $k = 0, 1, 2, ..$ and $-\cos^{-1} c + 2k\pi$, with $k = 1, 2, 3, ...$ as shown in Figure 11.23. Note that we have plotted contours for $c = 0, 0.2, 0.4, 0.6, 0.8, 1$.

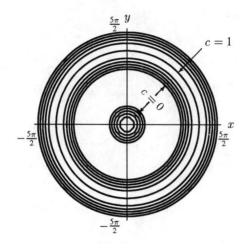

Figure 11.23

13.   We'll set $z = 4$ at the peak.

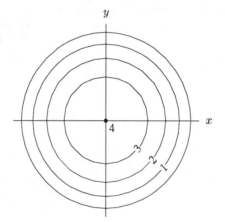

Figure 11.24

17.

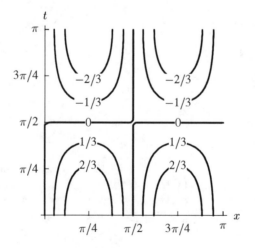

Figure 11.25

21. (a) The TMS map of an eye of constant curvature will have only one color, with no contour lines dividing the map.

     (b) The contour lines are circles, because the cross-section is the same in every direction. The largest curvature is in the center.

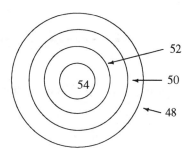

25. (a) To find the level curves, we let T be a constant.

$$T = 100 - x^2 - y^2$$
$$x^2 + y^2 = 100 - T,$$

which is an equation for a circle of radius $\sqrt{100 - T}$ centered at the origin. At $T = 100°$, we have a circle of radius 0 (a point). At $T = 75°$, we have a circle of radius 5. At $T = 50°$, we have a circle of radius $5\sqrt{2}$. At $T = 25°$, we have a circle of radius $5\sqrt{3}$. At $T = 0°$, we have a circle of radius 10.

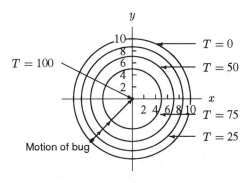

Figure 11.26

     (b) No matter where we put the bug, it should go straight toward the origin—the hottest point on the xy-plane. Its direction of motion is perpendicular to the tangent lines of the level curves, as can be seen in Figure 11.26.

29. Suppose P_0 is the production given by L_0 and K_0, so that

$$P_0 = f(L_0, K_0) = cL_0^\alpha K_0^\beta.$$

We want to know what happens to production if L_0 is increased to $2L_0$ and K_0 is increased to $2K_0$:

$$\begin{aligned}
P &= f(2L_0, 2K_0) \\
&= c(2L_0)^\alpha (2K_0)^\beta \\
&= c2^\alpha L_0^\alpha 2^\beta K_0^\beta \\
&= 2^{\alpha+\beta} cL_0^\alpha K_0^\beta \\
&= 2^{\alpha+\beta} P_0.
\end{aligned}$$

Thus, doubling L and K has the effect of multiplying P by $2^{\alpha+\beta}$. Notice that if $\alpha + \beta > 1$, then $2^{\alpha+\beta} > 2$, if $\alpha + \beta = 1$, then $2^{\alpha+\beta} = 2$, and if $\alpha + \beta < 1$, then $2^{\alpha+\beta} < 2$. Thus, $\alpha + \beta > 1$ gives increasing returns to scale, $\alpha + \beta = 1$ gives constant returns to scale, and $\alpha + \beta < 1$ gives decreasing returns to scale.

Solutions for Section 11.5

1. (a) Since z is a linear function of x and y with slope 2 in the x-direction, and slope 3 in the y-direction, we have:

$$z = 2x + 3y + c$$

We can write an equation for changes in z in terms of changes in x and y:

$$\Delta z = (2(x + \Delta x) + 3(y + \Delta y) + c) - (2x + 3y + c)$$
$$= 2\Delta x + 3\Delta y$$

Since $\Delta x = 0.5$ and $\Delta y = -0.2$, we have

$$\Delta z = 2(0.5) + 3(-0.2) = 0.4$$

So a 0.5 change in x and a -0.2 change in y produces a 0.4 change in z.

(b) As we know that $z = 2$ when $x = 5$ and $y = 7$, the value of z when $x = 4.9$ and $y = 7.2$ will be

$$z = 2 + \Delta z = 2 + 2\Delta x + 3\Delta y$$

where Δz is the change in z when x changes from 4.9 to 5 and y changes from 7.2 to 7. We have $\Delta x = 4.9 - 5 = -0.1$ and $\Delta y = 7.2 - 7 = 0.2$. Therefore, when $x = 4.9$ and $y = 7.2$, we have

$$z = 2 + 2 \cdot (-0.1) + 5 \cdot 0.2 = 2.4$$

5. When $y = 0$, $c + mx = 3x + 4$, so $c = 4$, $m = 3$. Thus, when $x = 0$, we have $4 + ny = y + 4$, so $n = 1$. Thus, $z = 4 + 3x + y$.

9. For each column in the table, we find that as x increases by 1, $f(x, y)$ increases by 2, so the x slope is 2. For each row in the table, we find that as y increases by 1, $f(x, y)$ decreases by 0.5, so the y slope is -0.5. So the function has the form $f(x, y) = 2x - 0.5y + c$. Also note that $f(0, 0) = 1$, so $c = 1$. Therefore, the function is $f(x, y) = 2x - 0.5y + 1$.

13. In the diagram the contours correspond to values of the function that are 2 units apart, i.e., there are contours for $-2, 0, 2$, etc. Note that moving two units in the y direction we cross three contours; i.e., a change of 2 in y changes the function by 6, so the y slope is 3. Similarly, a move of 1 in the positive x direction crosses one contour line and changes the function by -2; so the x slope is -2. Hence $f(x, y) = c - 2x + 3y$. We see from the diagram that $f(0, 1) = 6$. Solving for c gives $c = 3$. Therefore the function is $f(x, y) = 3 - 2x + 3y$.

17.

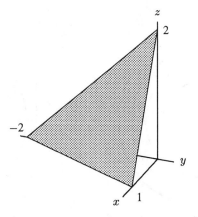

Figure 11.27

Solutions for Section 11.6

1. (a) On the surface, in the corner where the hot water enters, the temperature is highest. As we move further away from this corner, the temperature of the water decreases. In the corner furthest from the hot water's entry point, the water is the coolest. A possible contour diagram is shown in Figure 11.28. Temperatures are in C°.

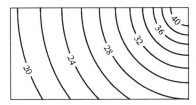

Figure 11.28: Surface temperatures

(b) One meter below the surface, the water is cooler than that at the surface, but still it is warmest in the corner where the hot water enters and gets cooler further away. The contour diagram is similar, but the temperature is lower because of the water's depth. (See Figure 11.29.)

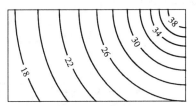

Figure 11.29: Temperatures at depth of one meter

5.

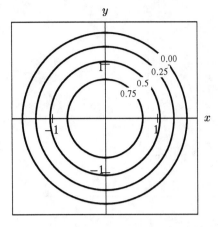

Figure 11.30: Contour diagram when $t = 0$

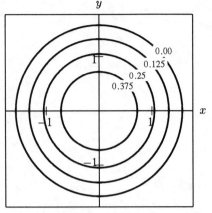

Figure 11.31: Contour diagram when $t = \pi/3$

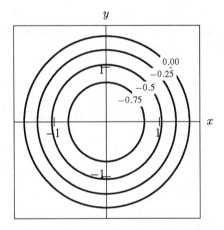

Figure 11.32: Contour diagram when $t = \pi$

9. Suppose $w = f(x, y, z) = ax + by + cz + d$.

We get b, the y slope, from the table with $z = 4$. We have $b = \Delta w/\Delta y = (9 - 14)/(5 - 3) = -5/2$.

Since the y slope is $-5/2$, the bottom left entry in the $z = 1$ table is obtained from the 8 in the same table by adding 5 to give 13. Thus $f(1, 3, 1) = 13$. Thus the x slope is given by $a = \Delta w/\Delta x = (13-4)/(1-0) = 9$.

We get c, the z slope, by comparing the values of w when $x = 1$, $y = 5$. Then $c = \Delta w/\Delta z = (9 - 8)/(4 - 1) = 1/3$.

Now use the fact that $a = 9, b = -5/2, c = 1/3$ and

$$f(0, 3, 1) = 9(0) - \frac{5}{2}(3) + \frac{1}{3}(1) + d = 4$$

so $d = 67/6$. Thus

$$f(x, y, z) = 9x - \frac{5}{2}y + \frac{1}{3}z + \frac{67}{6}$$

13. A hyperboloid of two sheets.

17. An ellipsoid.

Solutions for Section 11.7

1. Let us suppose that (x, y) tends to $(0,0)$ along the curve $y = kx^2$, where $k \neq -1$. We get

$$f(x, y) = f(x, kx^2) = \frac{x^2}{x^2 + kx^2} = \frac{1}{1 + k}.$$

Therefore:

$$\lim_{x \to 0} f(x, kx^2) = \frac{1}{1 + k}$$

and so for $k = 0$ we get

$$\lim_{\substack{(x,y) \to (0,0) \\ y=0}} f(x, y) = 1$$

and for $k = 1$

$$\lim_{\substack{(x,y) \to (0,0) \\ y=x^2}} f(x, y) = \frac{1}{2}.$$

Thus no matter how close they are to the origin, there will be points (x, y) where the value $f(x, y)$ is close to 1 and points (x, y) where $f(x, y)$ is close to $\frac{1}{2}$. So the limit:

$$\lim_{(x,y) \to (0,0)} f(x, y)$$

doesn't exist.

5. Since the composition of continuous functions is continuous, the function f is continuous at $(0,0)$. We have:

$$\lim_{(x,y) \to (0,0)} f(x, y) = \lim_{(x,y) \to (0,0)} (x^2 + y^2) = 0 + 0 = 0.$$

9. We want to compute

$$\lim_{(x,y) \to (0,0)} f(x, y) = \lim_{(x,y) \to (0,0)} \frac{\sin(x^2 + y^2)}{x^2 + y^2}.$$

As $r = \sqrt{x^2 + y^2}$ is the distance from (x, y) to $(0,0)$ we have that $(x, y) \to (0,0)$ is equivalent to $r \to 0$. Hence the limit becomes:

$$\lim_{(x,y) \to (0,0)} f(x, y) = \lim_{r \to 0} \frac{\sin r^2}{r^2} = 1.$$

13. A contour of g is the curve defined by the equation $g(x, y) = c$, where c is a constant.

$$\frac{x^2}{x^2 + y^2} = c \quad \text{or} \quad x^2 = cx^2 + cy^2 \text{ or again}$$
$$(1 - c)x^2 = cy^2.$$

Thus if $c < 0$ or $c > 1$ the only point which satisfies the above equation is $(0, 0)$ but this is not in the domain of g. For these values of c there are no contours.

If $c = 0$ we get $x = 0$ which is the y-axis without the origin (hence two rays of undefined slope).
If $c = 1$ we get $y = 0$ which is the x-axis without the origin (hence two rays of slope 0).
Finally, if $0 < c < 1$, we get two lines without the origin, namely:

$$y = \pm\sqrt{\frac{1 - c}{c}}x, \quad x \neq 0.$$

Therefore we get four rays of slopes $\sqrt{\frac{1-c}{c}}$ and $-\sqrt{\frac{1-c}{c}}$ respectively.

Solutions for Chapter 11 Review

1. These conditions describe a line parallel to the z-axis which passes through the xy-plane at $(2, 1, 0)$.

5. When h is fixed, say $h = 1$, then
$$V = f(r, 1) = \pi r^2 1 = \pi r^2$$

Similarly,
$$f\left(r, \frac{2}{3}\right) = \frac{4}{9}\pi r^2 \quad \text{and} \quad f\left(r, \frac{1}{3}\right) = \frac{\pi}{9}r^2$$

When r is fixed, say $r = 1$, then
$$f(1, h) = \pi(1)^2 h = \pi h$$

Similarly,
$$f(2, h) = 4\pi \quad \text{and} \quad f(3, h) = 9\pi h.$$

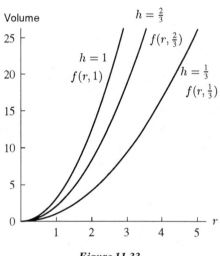

Figure 11.33

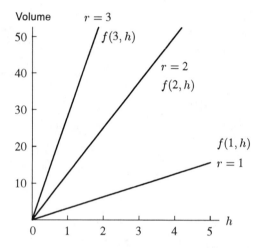

Figure 11.34

9.

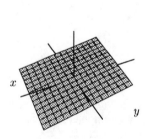

One possible equation: $x + y + z = 1$.

13. Might be true. The function $z = x^2 - y^2 + 1$ has this property. The level curve $z = 1$ is the lines $y = x$ and $y = -x$.

17. Contours are lines of the form $3x - 5y + 1 = c$ as shown in Figure 11.35. Note that for the regions of x and y given, the c values range from $-15 < c < 17$.

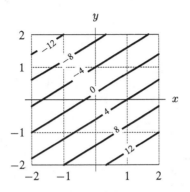

Figure 11.35

21. The line $t = 5$ crosses the contour $H = 80$ at about $x = 4$; this means that $H(4, 5) \approx 80$, and so the point $(4, 80)$ is on the graph of the one-variable function $y = H(x, 5)$. Each time the line crosses a contour, we can plot another point on the graph of $H(x, 5)$, and thus get a sketch of the graph. See Figure 11.36. Each data point obtained from the contour map has been indicated by a dot on the graph. The graph of $H(x, 20)$ was obtained in a similar way.

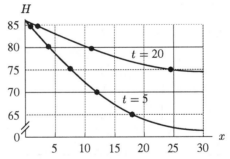

Figure 11.36: Graph of $H(x, 5)$ and $H(x, 20)$:
heat as a function of distance from the heater at
$t = 5$ and $t = 20$ minutes

These two graphs describe the temperature at different positions as a function of x for $t = 5$ and $t = 20$. Notice that the graph of $H(x, 5)$ descends more steeply than the graph of $H(x, 20)$; this is because the contours are quite close together along the line $t = 5$, whereas they are more spread out along the line $t = 20$. In practical terms the shape of the graph of $H(x, 5)$ tells us that the temperature drops quickly as you move away from the heater, which makes sense, since the heater was turned on just five minutes ago. On the other hand, the graph of $H(x, 20)$ descends more slowly, which makes sense, because the heater has been on for 20 minutes and the heat has had time to diffuse throughout the room.

25. (a) Let x = distance (microns) from center of waveguide, t = time (nanoseconds) as shown in the problem, and I = intensity of light as marked on the given level curves.

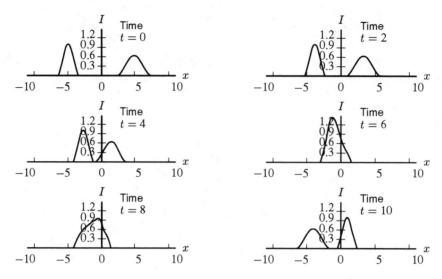

Figure 11.37

(b)   Two waves would start out at opposite ends of the screen. The wave on the left would be slightly taller and narrower than the wave on the right. The waves would move toward one another, the wave on the right moving a little faster. They would meet to the left of the center and appear to merge, becoming taller. They would then proceed in the directions they were initially going, ultimately leaving the screen on the side opposite to where they began.

(c)   Let x = distance (microns), t = time (nanoseconds), and I = intensity.

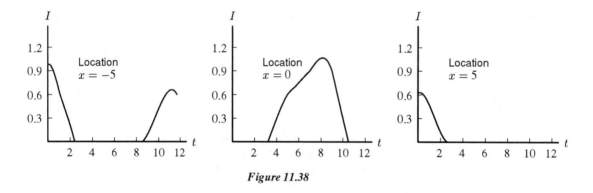

Figure 11.38

(d)   Two pulses of light are traveling down a wave-guide toward one another. They meet in the center and, as they pass through one another, appear brighter. They then continue along in the wave-guide in the directions they were going.

CHAPTER TWELVE

Solutions for Section 12.1

1.

$$\vec{p} = 2\vec{w}, \quad \vec{q} = -\vec{u}, \quad \vec{r} = \vec{w} + \vec{u} = \vec{u} + \vec{w},$$
$$\vec{s} = \vec{p} + \vec{q} = 2\vec{w} - \vec{u}, \quad \vec{t} = \vec{u} - \vec{w}$$

5. $5\vec{b} = 5(-3\vec{i} + 5\vec{j} + 4\vec{k}) = -15\vec{i} + 25\vec{j} + 20\vec{k}$.

9. $4\vec{i} + 2\vec{j} - 3\vec{i} + \vec{j} = \vec{i} + 3\vec{j}$

13. $\|\vec{v}\| = \sqrt{1^2 + (-1)^2 + 3^2} = \sqrt{11}$.

17. We get displacement by subtracting the coordinates of the origin $(0,0,0)$ from the coordinates of the cat $(1,4,0)$, giving
Displacement $= (1 - 0)\vec{i} + (4 - 0)\vec{j} + (0 - 0)\vec{k} = \vec{i} + 4\vec{j}$.

21.

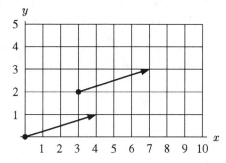

Figure 12.1: $\vec{v}$

25. $\vec{a} = \vec{b} = \vec{c} = 3\vec{k}, \quad \vec{d} = 2\vec{i} + 3\vec{k}, \quad \vec{e} = \vec{j}, \quad \vec{f} = -2\vec{i}$

29. (a) The displacement from P to Q is given by

$$\vec{PQ} = (4\vec{i} + 6\vec{j}) - (\vec{i} + 2\vec{j}) = 3\vec{i} + 4\vec{j}.$$

Since

$$\|\vec{PQ}\| = \sqrt{3^2 + 4^2} = 5,$$

a unit vector $\vec{u}$ in the direction of $\vec{PQ}$ is given by

$$\vec{u} = \frac{1}{5}\vec{PQ} = \frac{1}{5}(3\vec{i} + 4\vec{j}) = \frac{3}{5}\vec{i} + \frac{4}{5}\vec{j}.$$

(b) A vector of length 10 pointing in the same direction is given by

$$10\vec{u} = 10(\frac{3}{5}\vec{i} + \frac{4}{5}\vec{j}) = 6\vec{i} + 8\vec{j}.$$

33.

Figure 12.2

Solutions for Section 12.2

1.   Scalar

5.   (a)   If the car is going east, it is going solely in the positive x direction, so its velocity vector is $50\vec{i}$.
   (b)   If the car is going south, it is going solely in the negative y direction, so its velocity vector is $-50\vec{j}$.
   (c)   If the car is going southeast, the angle between the x-axis and the velocity vector is $-45°$. Therefore

$$\text{velocity vector} = 50\cos(-45°)\vec{i} + 50\sin(-45°)\vec{j}$$
$$= 25\sqrt{2}\vec{i} - 25\sqrt{2}\vec{j}.$$

   (d)   If the car is going northwest, the velocity vector is at a $45°$ angle to the y-axis, which is $135°$ from the x-axis. Therefore:

$$\text{velocity vector} = 50(\cos 135°)\vec{i} + 50(\sin 135°)\vec{j} = -25\sqrt{2}\vec{i} + 25\sqrt{2}\vec{j}.$$

9.   At the point P, the velocity of the car is changing the quickest; not in magnitude, but in direction only. The acceleration vector is therefore the longest at this point. The direction of the vector is directed in towards the center of the track because the difference in velocity vectors at nearby points is a vector pointing toward the center.

13.   Suppose $\vec{u}$ represents the velocity of the plane relative to the air and $\vec{w}$ represents the velocity of the wind. We can add these two vectors by adding their components. Suppose north is in the y-direction and east is the x-direction. The vector representing the airplane's velocity makes an angle of $45°$ with north; the components of $\vec{u}$ are

$$\vec{u} = 700\sin 45°\vec{i} + 700\cos 45°\vec{j} \approx 495\vec{i} + 495\vec{j}.$$

Since the wind is blowing from the west, $\vec{w} = 60\vec{i}$. By adding these we get a resultant vector $\vec{v} = 555\vec{i} + 495\vec{j}$. The direction relative to the north is the angle θ shown in Figure 12.3 given by

$$\theta = \tan^{-1}\frac{x}{y} = \tan^{-1}\frac{555}{495}$$
$$\approx 48.3°$$

The magnitude of the velocity is

$$\|\vec{v}\| = \sqrt{495^2 + 555^2} = \sqrt{553{,}050}$$
$$= 744 \text{ km/hr}.$$

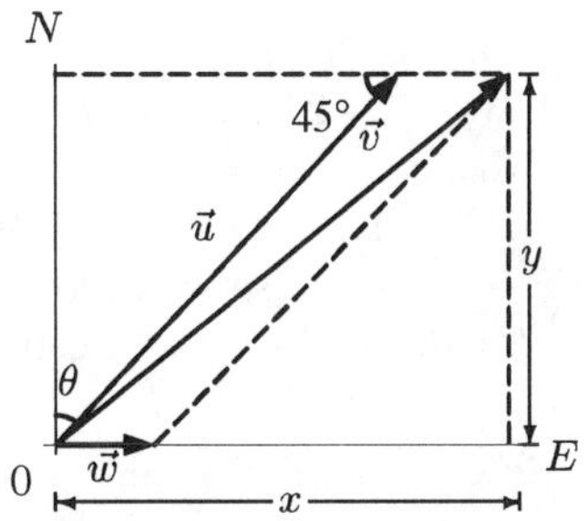

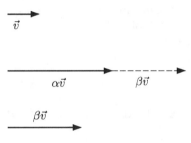

Figure 12.3: Note that θ is the angle between north and the vector $\vec{v}$

17.

Figure 12.4

The vectors $\vec{v}$, $\alpha\vec{v}$ and $\beta\vec{v}$ are all parallel. Figure 12.4 shows them with α, $\beta > 0$, so all the vectors are in the same direction. Notice that $\alpha\vec{v}$ is a vector α times as long as $\vec{v}$ and $\beta\vec{v}$ is β times as long as $\vec{v}$. Therefore $\alpha\vec{v} + \beta\vec{v}$ is a vector $(\alpha + \beta)$ times as long as $\vec{v}$, and in the same direction. Thus,

$$\alpha\vec{v} + \beta\vec{v} = (\alpha + \beta)\vec{v}.$$

21. Since the zero vector has zero length, adding it to $\vec{v}$ has no effect.

Solutions for Section 12.3

1. $\vec{c} \cdot \vec{y} = (\vec{i} + 6\vec{j}) \cdot (4\vec{i} - 7\vec{j}) = (1)(4) + (6)(-7) = 4 - 42 = -38$.

5. Since $\vec{a} \cdot \vec{y}$ and $\vec{c} \cdot \vec{z}$ are both scalars, the answer to this equation is the product of two numbers and therefore a number. We have

$$\vec{a} \cdot \vec{y} = (2\vec{j} + \vec{k}) \cdot (4\vec{i} - 7\vec{j}) = 0(4) + 2(-7) + 1(0) = -14$$

$$\vec{c} \cdot \vec{z} = (\vec{i} + 6\vec{j}) \cdot (\vec{i} - 3\vec{j} - \vec{k}) = 1(1) + 6(-3) + 0(-1) = -17$$

Thus,

$$(\vec{a} \cdot \vec{y})(\vec{c} \cdot \vec{z}) = 238$$

9. In general, $\vec{u}$ and $\vec{v}$ are perpendicular when $\vec{u} \cdot \vec{v} = 0$.
   In this case, $\vec{u} \cdot \vec{v} = (t\vec{i} - \vec{j} + \vec{k}) \cdot (t\vec{i} + t\vec{j} - 2\vec{k}) = t^2 - t - 2$.
   This is zero when $t^2 - t - 2 = 0$, i.e. when $(t - 2)(t + 1) = 0$, so $t = 2$ or -1.
   In general, $\vec{u}$ and $\vec{v}$ are parallel if and only if $\vec{v} = \alpha\vec{u}$ for some real number α.
   Thus we need $\alpha t\vec{i} - \alpha\vec{j} + \alpha\vec{k} = t\vec{i} + t\vec{j} - 2\vec{k}$, so we need $\alpha t = t$, and $-\alpha = t$, and $\alpha = -2$. But if $\alpha = -2$, we can't have $\alpha t = t$ unless $t = 0$, and if $t = 0$, we can't have $-\alpha = t$, so there are no values of t for which $\vec{u}$ and $\vec{v}$ are parallel.

13. Since a normal vector of the plane is $\vec{n} = -\vec{i} + 2\vec{j} + \vec{k}$, an equation for the plane is

$$-x + 2y + z = -1 + 2 \cdot 0 + 2 = 1$$
$$-x + 2y + z = 1.$$

17. Two planes are parallel if their normal vectors are parallel. Since the plane $3x + y + z = 4$ has normal vector $\vec{n} = 3\vec{i} + \vec{j} + \vec{k}$, the plane we are looking for has the same normal vector and passes through the point $(-2, 3, 2)$. Thus, it has the equation

$$3x + y + z = 3 \cdot (-2) + 3 + 2 = -1.$$

21. The angle between two planes is equal to the angle between the normal vectors of the two planes. A normal vector to the plane $5(x - 1) + 3(y + 2) + 2z = 0$ is

$$\vec{n}_1 = 5\vec{i} + 3\vec{j} + 2\vec{k},$$

and a normal vector to the plane $x + 3(y - 1) + 2(z + 4) = 0$ is

$$\vec{n}_2 = \vec{i} + 3\vec{j} + 2\vec{k}.$$

Since $\vec{n}_1 \cdot \vec{n}_2 = \|\vec{n}_1\|\|\vec{n}_2\|\cos\theta$, then

$$\cos\theta = \frac{\vec{n}_1 \cdot \vec{n}_2}{\|\vec{n}_1\|\|\vec{n}_2\|} = \frac{(5\vec{i} + 3\vec{j} + 2\vec{k}) \cdot (\vec{i} + 3\vec{j} + 2\vec{k})}{\sqrt{5^2 + 3^2 + 2^2}\sqrt{1^2 + 3^2 + 2^2}}$$
$$= \frac{18}{\sqrt{532}} = 0.78$$

Hence, $\theta \approx 38.7°$.

25. We have

$$\|\vec{a}_2\| = \sqrt{0.10^2 + 0.08^2 + 0.12^2 + 0.69^2} = 0.7120$$
$$\|\vec{a}_3\| = \sqrt{0.20^2 + 0.06^2 + 0.06^2 + 0.66^2} = 0.6948$$
$$\|\vec{a}_4\| = \sqrt{0.22^2 + 0.00^2 + 0.20^2 + 0.57^2} = 0.6429$$

$$\vec{a}_2 \cdot \vec{a}_3 = 0.10 \cdot 0.20 + 0.08 \cdot 0.06 + 0.12 \cdot 0.06 + 0.69 \cdot 0.66 = 0.4874$$
$$\vec{a}_3 \cdot \vec{a}_4 = 0.20 \cdot 0.22 + 0.06 \cdot 0.00 + 0.06 \cdot 0.20 + 0.66 \cdot 0.57 = 0.4322$$

The distance between the English and the Bantus is given by θ where

$$\cos\theta = \frac{\vec{a}_2 \cdot \vec{a}_3}{\|\vec{a}_2\|\|\vec{a}_3\|} = \frac{0.4874}{(0.7120)(0.6948)} \approx 0.9852$$

so $\theta \approx 9.9°$.

The distance between the English and the Koreans is given by ϕ where

$$\cos\phi = \frac{\vec{a}_3 \cdot \vec{a}_4}{\|\vec{a}_3\|\|\vec{a}_4\|} = \frac{0.4322}{(0.6948)(0.6429)} \approx 0.9676$$

so $\phi \approx 14.6°$. Hence the English are genetically closer to the Bantus than to the Koreans.

29. Since $\vec{u} \cdot \vec{w} = \vec{v} \cdot \vec{w}$, $(\vec{u} - \vec{v}) \cdot \vec{w} = 0$. This equality holds for any $\vec{w}$, so we can take $\vec{w} = \vec{u} - \vec{v}$. This gives

$$\|\vec{u} - \vec{v}\|^2 = (\vec{u} - \vec{v}) \cdot (\vec{u} - \vec{v}) = 0,$$

that is,

$$\|\vec{u} - \vec{v}\| = 0.$$

This implies $\vec{u} - \vec{v} = 0$, that is, $\vec{u} = \vec{v}$.

33. We substitute $\vec{u} = u_1\vec{i} + u_2\vec{j} + u_3\vec{k}$ and by the result of Problem 30, we expand as follows:

$$
\begin{aligned}
(\vec{u} \cdot \vec{v})_{\text{geom}} &= (u_1\vec{i} + u_2\vec{j} + u_3\vec{k}) \cdot \vec{v} \\
&= (u_1\vec{i}) \cdot \vec{v} + (u_2\vec{j}) \cdot \vec{v} + (u_3\vec{k}) \cdot \vec{v}
\end{aligned}
$$

where all the dot products are defined geometrically By the result of Problem 31 we can write

$$(\vec{u} \cdot \vec{v})_{\text{geom}} = u_1(\vec{i} \cdot \vec{v})_{\text{geom}} + u_2(\vec{j} \cdot \vec{v})_{\text{geom}} + u_3(\vec{k} \cdot \vec{v})_{\text{geom}}.$$

Now substitute $\vec{v} = v_1\vec{i} + v_2\vec{j} + v_3\vec{k}$ and expand, again using Problem 30 and the geometric definition of the dot product:

$$
\begin{aligned}
(\vec{u} \cdot \vec{v})_{\text{geom}} &= u_1 \left(\vec{i} \cdot (v_1\vec{i} + v_2\vec{j} + v_3\vec{k})\right)_{\text{geom}} \\
&\quad + u_2 \left(\vec{j} \cdot (v_1\vec{i} + v_2\vec{j} + v_3\vec{k})\right)_{\text{geom}} \\
&\quad + u_3 \left(\vec{k} \cdot (v_1\vec{i} + v_2\vec{j} + v_3\vec{k})\right)_{\text{geom}} \\
&= u_1v_1(\vec{i} \cdot \vec{i})_{\text{geom}} + u_1v_2(\vec{i} \cdot \vec{j})_{\text{geom}} + u_1v_3(\vec{i} \cdot \vec{k})_{\text{geom}} \\
&\quad + u_2v_1(\vec{i} \cdot \vec{i})_{\text{geom}} + u_2v_2(\vec{i} \cdot \vec{j})_{\text{geom}} + u_2v_3(\vec{i} \cdot \vec{k})_{\text{geom}} \\
&\quad + u_3v_1(\vec{i} \cdot \vec{i})_{\text{geom}} + u_3v_2(\vec{i} \cdot \vec{j})_{\text{geom}} + u_3v_3(\vec{i} \cdot \vec{k})_{\text{geom}}
\end{aligned}
$$

The geometric definition of the dot product shows that

$$
\begin{aligned}
\vec{i} \cdot \vec{i} &= \|\vec{i}\|\|\vec{i}\|\cos 0 = 1 \\
\vec{i} \cdot \vec{j} &= \|\vec{i}\|\|\vec{j}\|\cos\frac{\pi}{2} = 0.
\end{aligned}
$$

Similarly $\vec{j} \cdot \vec{j} = \vec{k} \cdot \vec{k} = 1$ and $\vec{i} \cdot \vec{k} = \vec{j} \cdot \vec{k} = 0$. Thus, the expression for $(\vec{u} \cdot \vec{v})_{\text{geom}}$ becomes

$$
\begin{aligned}
(\vec{u} \cdot \vec{v})_{\text{geom}} &= u_1v_1(1) + u_1v_2(0) + u_1v_3(0) \\
&\quad + u_2v_1(0) + u_2v_2(1) + u_2v_3(0) \\
&\quad + u_3v_1(0) + u_3v_2(0) + u_3v_3(1) \\
&= u_1v_1 + u_2v_2 + u_3v_3.
\end{aligned}
$$

Solutions for Section 12.4

1. $\vec{k} \times \vec{j} = -\vec{i}$ (remember $\vec{i}, \vec{j}, \vec{k}$ are unit vectors along the axes, and you must use the right hand rule.)

5. $\vec{a} = \vec{i} + \vec{j} + \vec{k}$, and $\vec{b} = \vec{i} + \vec{j} - \vec{k}$

$$\vec{a} \times \vec{b} = \begin{vmatrix} \vec{i} & \vec{j} & \vec{k} \\ 1 & 1 & 1 \\ 1 & 1 & -1 \end{vmatrix} = -2\vec{i} + 2\vec{j}$$

9. The magnitude of $\vec{a} \times \vec{b}$ is given by

$$\|\vec{a} \times \vec{b}\| = \|\vec{a}\| \|\vec{b}\| \sin \theta = (3 \cdot 2) \sin \theta = 6 \sin \theta.$$

Therefore, the maximum possible value of $\|\vec{a} \times \vec{b}\|$ occurs when $\sin \theta = 1$. This occurs when $\theta = \pi/2$; that is, when $\vec{a}$ and $\vec{b}$ are perpendicular. The maximum value of $\|\vec{a} \times \vec{b}\|$ is 6. The minimum value of $\|\vec{a} \times \vec{b}\|$ occurs when $\sin \theta = 0$ so $\theta = 0$ or π; that is, when $\vec{a}$ and $\vec{b}$ are parallel. Then $\|\vec{a} \times \vec{b}\|$ is 0.

The direction of $\vec{a} \times \vec{b}$ will be along the positive z-axis when $\vec{b}$ is in the first or second quadrant and along the negative z-axis when $\vec{b}$ is in the third or fourth quadrant. See Figure 12.5.

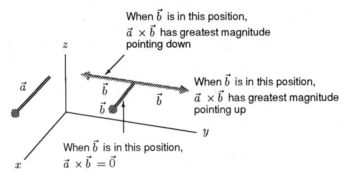

Figure 12.5: A fixed vector $\vec{a}$ and a rotating vector $\vec{b}$

13. (a) If we let $\overrightarrow{PQ}$ in Figure 12.6 be the vector from point P to point Q and $\overrightarrow{PR}$ be the vector from P to R, then

$$\overrightarrow{PQ} = -\vec{i} + 2\vec{k}$$
$$\overrightarrow{PR} = 2\vec{i} - \vec{k},$$

then the area of the parallelogram determined by $\overrightarrow{PQ}$ and $\overrightarrow{PR}$ is:

$$\begin{array}{c} \text{Area of} \\ \text{parallelogram} \end{array} = \|\overrightarrow{PQ} \times \overrightarrow{PR}\| = \left\| \begin{vmatrix} \vec{i} & \vec{j} & \vec{k} \\ -1 & 0 & 2 \\ 2 & 0 & -1 \end{vmatrix} \right\| = \|3\vec{j}\| = 3.$$

Thus, the area of the triangle PQR is

$$\left(\begin{array}{c} \text{Area of} \\ \text{triangle} \end{array} \right) = \frac{1}{2} \left(\begin{array}{c} \text{Area of} \\ \text{parallelogram} \end{array} \right) = \frac{3}{2} = 1.5.$$

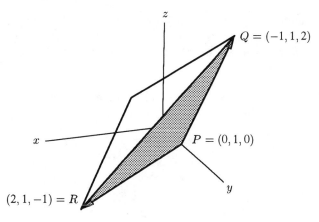

Figure 12.6

(b) Since $\vec{n} = \overrightarrow{PQ} \times \overrightarrow{PR}$ is perpendicular to the plane PQR, and from above, we have $\vec{n} = 3\vec{j}$, the equation of the plane has the form $3y = C$. At the point $(0, 1, 0)$ we get $3 = C$, therefore $3y = 3$, i.e., $y = 1$.

17. First let
$$\vec{a} = a_1\vec{i} + a_2\vec{j} + a_3\vec{k} \quad \vec{b} = b_1\vec{i} + b_2\vec{j} + b_3\vec{k} \quad \vec{c} = c_1\vec{i} + c_2\vec{j} + c_3\vec{k}$$
so $\vec{b} + \vec{c} = (b_1 + c_1)\vec{i} + (b_2 + c_2)\vec{j} + (b_3 + c_3)\vec{k}$. Now, using the general formula for cross products, we have:

$$\vec{a} \times (\vec{b} + \vec{c})$$
$$= [a_2(b_3 + c_3) - a_3(b_2 + c_2)]\vec{i} + [a_3(b_1 + c_1) - a_1(b_3 + c_3)]\vec{j} + [a_1(b_2 + c_2) - a_2(b_1 + c_1)]\vec{k}$$
$$= (a_2b_3 + a_2c_3 - a_3b_2 - a_3c_2)\vec{i} + (a_3b_1 + a_3c_1 - a_1b_3 - a_1c_3)\vec{j}$$
$$+ (a_1b_2 + a_1c_2 - a_2b_1 - a_2c_1)\vec{k}$$
$$= (a_2b_3 - a_3b_2)\vec{i} + (a_2c_3 - a_3c_2)\vec{i} + (a_3b_1 - a_1b_3)\vec{j} + (a_3c_1 - a_1c_3)\vec{j}$$
$$+ (a_1b_2 - a_2b_1)\vec{k} + (a_1c_2 - a_2c_1)\vec{k}$$
$$= (a_2b_3 - a_3b_2)\vec{i} + (a_3b_1 - a_1b_3)\vec{j} + (a_1b_2 - a_2b_1)\vec{k} + (a_2c_3 - a_3c_2)\vec{i} + (a_3c_1 - a_1c_3)\vec{j}$$
$$+ (a_1c_2 - a_2c_1)\vec{k}$$
$$= (\vec{a} \times \vec{b}) + (\vec{a} \times \vec{c})$$

Thus, $\vec{a} \times (\vec{b} + \vec{c}) = \vec{a} \times \vec{b} + \vec{a} \times \vec{c}$.

21. If $\lambda = 0$, then all three cross products are $\vec{0}$, since the cross product of the zero vector with any other vector is always 0.

    If $\lambda > 0$, then $\lambda\vec{v}$ and $\vec{v}$ are in the same direction and $\vec{w}$ and $\lambda\vec{w}$ are in the same direction. Therefore the unit normal vector $\vec{n}$ is the same in all three cases. In addition, the angles between $\lambda\vec{v}$ and $\vec{w}$, and between $\vec{v}$ and $\vec{w}$, and between $\vec{v}$ and $\lambda\vec{w}$ are all θ. Thus,

$$(\lambda\vec{v}) \times \vec{w} = \|\lambda\vec{v}\|\|\vec{w}\| \sin\theta\vec{n}$$
$$= \lambda\|\vec{v}\|\|\vec{w}\| \sin\theta\vec{n}$$
$$= \lambda(\vec{v} \times \vec{w})$$
$$= \|\vec{v}\|\|\lambda\vec{w}\| \sin\theta\vec{n}$$
$$= \vec{v} \times (\lambda\vec{w})$$

If $\lambda < 0$, then $\lambda\vec{v}$ and $\vec{v}$ are in opposite directions, as are $\vec{w}$ and $\lambda\vec{w}$ in opposite directions. Therefore if $\vec{n}$ is the normal vector in the definition of $\vec{v} \times \vec{w}$, then the right-hand rule gives $-\vec{n}$ for $(\lambda\vec{v}) \times \vec{w}$ and $\vec{v} \times (\lambda\vec{w})$. In addition, if the angle between $\vec{v}$ and $\vec{w}$ is θ, then the angle between $\lambda\vec{v}$ and $\vec{w}$ and between $\vec{v}$ and $\lambda\vec{w}$ is $(\pi - \theta)$. Since if $\lambda < 0$, we have $|\lambda| = -\lambda$, so

$$
\begin{aligned}
(\lambda\vec{v}) \times \vec{w} &= \|\lambda\vec{v}\|\|\vec{w}\| \sin(\pi - \theta)(-\vec{n}) \\
&= |\lambda|\|\vec{v}\|\|\vec{w}\| \sin(\pi - \theta)(-\vec{n}) \\
&= -\lambda\|\vec{v}\|\|\vec{w}\| \sin\theta(-\vec{n}) \\
&= \lambda\|\vec{v}\|\|\vec{w}\| \sin\theta\vec{n} \\
&= \lambda(\vec{v} \times \vec{w}).
\end{aligned}
$$

Similarly,

$$
\begin{aligned}
\vec{v} \times (\lambda\vec{w}) &= \|\vec{v}\|\|\lambda\vec{w}\| \sin(\pi - \theta)(-\vec{n}) \\
&= -\lambda\|\vec{v}\|\|\vec{w}\| \sin\theta(-\vec{n}) \\
&= \lambda(\vec{v} \times \vec{w}).
\end{aligned}
$$

25. Problem 22 tells us that $(\vec{u} \times \vec{v}) \cdot \vec{w} = \vec{u} \cdot (\vec{v} \times \vec{w})$. Using this result on the triple product of $(\vec{a} + \vec{b}) \times \vec{c}$ with any vector $\vec{d}$ together with the fact that the dot product distributes over addition gives us:

$$
\begin{aligned}
[(\vec{a} + \vec{b}) \times \vec{c}] \cdot \vec{d} &= (\vec{a} + \vec{b}) \cdot (\vec{c} \times \vec{d}) \\
&= \vec{a} \cdot (\vec{c} \times \vec{d}) + \vec{b} \cdot (\vec{c} \times \vec{d}) \quad \text{(dot product is distributive)} \\
&= (\vec{a} \times \vec{c}) \cdot \vec{d} + (\vec{b} \times \vec{c}) \cdot \vec{d} \quad \text{(using Problem 22 again)} \\
&= [(\vec{a} \times \vec{c}) + (\vec{b} \times \vec{c})] \cdot \vec{d}. \quad \text{(dot product is distributive)}
\end{aligned}
$$

So, since $[(\vec{a} + \vec{b}) \times \vec{c}] \cdot \vec{d} = [(\vec{a} \times \vec{c}) + (\vec{b} \times \vec{c})] \cdot \vec{d}$, then

$$
[(\vec{a} + \vec{b}) \times \vec{c}] \cdot \vec{d} - [(\vec{a} \times \vec{c}) + (\vec{b} \times \vec{c})] \cdot \vec{d} = 0,
$$

Since the dot product is distributive, we have

$$
[((\vec{a} + \vec{b}) \times \vec{c}) - (\vec{a} \times \vec{c}) - (\vec{b} \times \vec{c})] \cdot \vec{d} = 0.
$$

Since this equation is true for all vectors $\vec{d}$, by letting

$$
\vec{d} = ((\vec{a} + \vec{b}) \times \vec{c}) - (\vec{a} \times \vec{c}) - (\vec{b} \times \vec{c}),
$$

we get

$$
\|(\vec{a} + \vec{b}) \times \vec{c} - \vec{a} \times \vec{c} - \vec{b} \times \vec{c}\|^2 = 0
$$

and hence

$$
(\vec{a} + \vec{b}) \times \vec{c} - (\vec{a} \times \vec{c}) - (\vec{b} \times \vec{c}) = \vec{0}.
$$

Thus

$$
(\vec{a} + \vec{b}) \times \vec{c} = (\vec{a} \times \vec{c}) + (\vec{b} \times \vec{c}).
$$

Solutions for Chapter 12 Review

1.

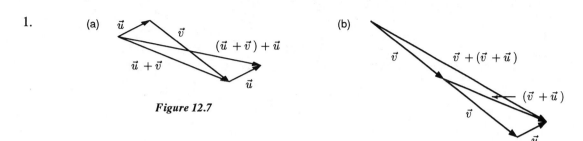

(a)

Figure 12.7

(b)

Figure 12.8

(c)

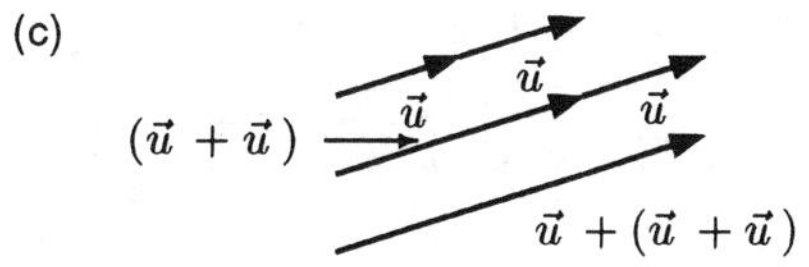

Figure 12.9

5.

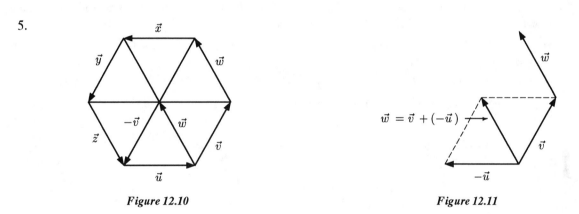

Figure 12.10

Figure 12.11

Break the hexagon up into 6 equilateral triangles, as shown in Figure 12.10.

Then $\vec{u} - \vec{v} + \vec{w} = \vec{0}$, so $\vec{w} = \vec{v} - \vec{u}$

Similarly, $\vec{x} = -\vec{u}$, $\vec{y} = -\vec{v}$, $\vec{z} = -\vec{w} = \vec{u} - \vec{v}$.

9.

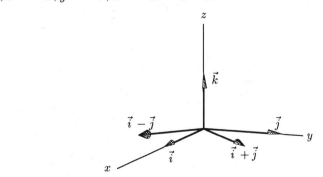

Figure 12.12

By definition, $(\vec{i} + \vec{j}) \times (\vec{i} - \vec{j})$ is in the direction of $-\vec{k}$. The magnitude is

$$\|\vec{i} + \vec{j}\| \cdot \|\vec{i} - \vec{j}\| \sin\frac{\pi}{4} = \sqrt{2} \cdot \sqrt{2} = 2.$$

So $(\vec{i} + \vec{j}) \times (\vec{i} - \vec{j}) = -2\vec{k}$. See Figure 12.12.

13. Let $\vec{r}_1$ be the displacement vector $\overrightarrow{PQ}$ and let $\vec{r}_2$ be the displacement vector $\overrightarrow{PR}$. Then

$$\vec{r}_1 = (1 + 2)\vec{i} + (3 - 2)\vec{j} + (-1 - 0)\vec{k} = 3\vec{i} + \vec{j} - \vec{k},$$
$$\vec{r}_2 = (-4 + 2)\vec{i} + (2 - 2)\vec{j} + (1 - 0)\vec{k} = -2\vec{i} + \vec{k},$$
$$\vec{r}_1 \times \vec{r}_2 = \begin{vmatrix} \vec{i} & \vec{j} & \vec{k} \\ 3 & 1 & -1 \\ -2 & 0 & 1 \end{vmatrix} = \vec{i} - (3 - 2)\vec{j} + 2\vec{k} = \vec{i} - \vec{j} + 2\vec{k}.$$

The area of the triangle $= \frac{1}{2}\|\vec{r}_1 \times \vec{r}_2\| = \frac{1}{2}\sqrt{1^2 + 1^2 + 2^2} = \frac{\sqrt{6}}{2}$.

17. (a) On the x-axis, $y = z = 0$, so $5x = 21$, giving $x = \frac{21}{5}$. So the only such point is $(\frac{21}{5}, 0, 0)$.
    (b) Other points are $(0, -21, 0)$, and $(0, 0, 3)$. There are many other possible answers.
    (c) $\vec{n} = 5\vec{i} - \vec{j} + 7\vec{k}$. It is the normal vector.
    (d) The vector between two points in the plane is parallel to the plane. Using the points from part (b), the vector $3\vec{k} - (-21\vec{j}) = 21\vec{j} + 3\vec{k}$ is parallel to the plane.

21. Writing $\vec{P} = (P_1, P_2, \cdots, P_{50})$ where P_i is the population of the i-th state, shows that $\vec{P}$ can be thought of as a vector with 50 components.

25. (a) Target A is at the point $(30, 0, 3)$; Target B is at the point $(20, 15, 0)$; Target C is the point $(12, 30, 8)$. You fire from the point $P = (0, 0, 5)$. The vectors to each of these targets are $\overrightarrow{PA} = 30\vec{i} - 2\vec{k}$, $\overrightarrow{PB} = 20\vec{i} + 15\vec{j} - 5\vec{k}$, $\overrightarrow{PC} = 12\vec{i} + 30\vec{j} + 3\vec{k}$.
    (b) You fire from the point $Q = (0, -1, 3)$, so $\overrightarrow{QA} = 30\vec{i} + \vec{j}$, $\overrightarrow{QB} = 20\vec{i} + 16\vec{j} - 3\vec{k}$, $\overrightarrow{QC} = 12\vec{i} + 31\vec{j} + 5\vec{k}$.

29. (a)

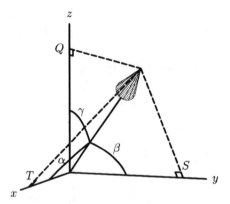

Figure 12.13

Suppose $\vec{v} = \overrightarrow{OP}$ as in Figure 12.13. The $\vec{i}$ component of $\overrightarrow{OP}$ is the projection of $\overrightarrow{OP}$ on the x-axis:

$$\overrightarrow{OT} = v\cos\alpha\vec{i}.$$

Similarly, the $\vec{j}$ and $\vec{k}$ components of $\overrightarrow{OP}$ are the projections of $\overrightarrow{OP}$ on the y-axis and the z-axis respectively. So:

$$\overrightarrow{OS} = v \cos \beta \vec{j}$$
$$\overrightarrow{OQ} = v \cos \gamma \vec{k}$$

Since $\vec{v} = \overrightarrow{OT} + \overrightarrow{OS} + \overrightarrow{OQ}$, we have

$$\vec{v} = v \cos \alpha \vec{i} + v \cos \beta \vec{j} + v \cos \gamma \vec{k}.$$

(b) Since

$$\begin{aligned} v^2 = \vec{v} \cdot \vec{v} &= (v \cos \alpha \vec{i} + v \cos \beta \vec{j} + v \cos \gamma \vec{k}) \cdot \\ &\quad (v \cos \alpha \vec{i} + v \cos \beta \vec{j} + v \cos \gamma \vec{k}) \\ &= v^2 (\cos^2 \alpha + \cos^2 \beta + \cos^2 \gamma) \end{aligned}$$

so

$$\cos^2 \alpha + \cos^2 \beta + \cos^2 \gamma = 1.$$

33. The displacement from $(1, 1, 1)$ to $(1, 4, 5)$ is

$$\vec{r_1} = (1 - 1)\vec{i} + (4 - 1)\vec{j} + (5 - 1)\vec{k} = 3\vec{j} + 4\vec{k}.$$

The displacement from $(-3, -2, 0)$ to $(1, 4, 5)$ is

$$\vec{r_2} = (1 + 3)\vec{i} + (4 + 2)\vec{j} + (5 - 0)\vec{k} = 4\vec{i} + 6\vec{j} + 5\vec{k}.$$

A normal vector is

$$\vec{n} = \vec{r_1} \times \vec{r_2} = \begin{vmatrix} \vec{i} & \vec{j} & \vec{k} \\ 0 & 3 & 4 \\ 4 & 6 & 5 \end{vmatrix} = (15 - 24)\vec{i} - (-16)\vec{j} + (-12)\vec{k} = -9\vec{i} + 16\vec{j} - 12\vec{k}.$$

The equation of the plane is

$$-9x + 16y - 12z = -9 \cdot 1 + 16 \cdot 1 - 12 \cdot 1 = -5$$
$$9x - 16y + 12z = 5.$$

We pick a point A on the plane, $A = (\frac{5}{9}, 0, 0)$ and let $P = (0, 0, 0)$. (See Figure 12.14.) Then $\overrightarrow{PA} = (5/9)\vec{i}$.

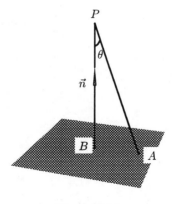

Figure 12.14

So the distance d from the point P to the plane is

$$d = \|\overrightarrow{PB}\| = \|\overrightarrow{PA}\| \cos\theta$$

$$= \frac{\overrightarrow{PA} \cdot \vec{n}}{\|\vec{n}\|} \quad \text{since } \overrightarrow{PA} \cdot \vec{n} = \|\overrightarrow{PA}\|\|\vec{n}\| \cos\theta)$$

$$= \left| \frac{\left(\frac{5}{9}\vec{i}\right) \cdot \left(-9\vec{i} + 16\vec{j} - 12\vec{k}\right)}{\sqrt{9^2 + 16^2 + 12^2}} \right|$$

$$= \frac{5}{\sqrt{481}} = 0.23.$$

CHAPTER THIRTEEN

1. If h is small, then

$$f_x(3,2) \approx \frac{f(3+h,2) - f(3,2)}{h}.$$

With $h = 0.01$, we find

$$f_x(3,2) \approx \frac{f(3.01,2) - f(3,2)}{0.01} = \frac{\frac{3.01^2}{(2+1)} - \frac{3^2}{(2+1)}}{0.01} = 2.00333.$$

With $h = 0.0001$, we get

$$f_x(3,2) \approx \frac{f(3.0001,2) - f(3,2)}{0.0001} = \frac{\frac{3.0001^2}{(2+1)} - \frac{3^2}{(2+1)}}{0.0001} = 2.0000333.$$

Since the difference quotient seems to be approaching 2 as h gets smaller, we conclude

$$f_x(3,2) \approx 2.$$

To estimate $f_y(3,2)$, we use

$$f_y(3,2) \approx \frac{f(3,2+h) - f(3,2)}{h}.$$

With $h = 0.01$, we get

$$f_y(3,2) \approx \frac{f(3,2.01) - f(3,2)}{0.01} = \frac{\frac{3^2}{(2.01+1)} - \frac{3^2}{(2+1)}}{0.01} = -0.99668.$$

With $h = 0.0001$, we get

$$f_y(3,2) \approx \frac{f(3,2.0001) - f(3,2)}{0.0001} = \frac{\frac{3^2}{(2.0001+1)} - \frac{3^2}{(2+1)}}{0.0001} = -0.9999667.$$

Thus, it seems that the difference quotient is approaching -1, so we estimate

$$f_y(3,2) \approx -1.$$

5. (a) An increase in the price of a new car will decrease the number of cars bought annually. Thus $\frac{\partial q_1}{\partial x} < 0$.
    Similarly, an increase in the price of gasoline will decrease the amount of gas sold, implying $\frac{\partial q_2}{\partial y} < 0$.
    (b) Since the demands for a car and gas complement each other, an increase in the price of gasoline will decrease the total number of cars bought. Thus $\frac{\partial q_1}{\partial y} < 0$. Similarly, we may expect $\frac{\partial q_2}{\partial x} < 0$.

9. (a) For points near the point $(0, 5, 3)$, moving in the positive x direction, the surface is sloping down and the function is decreasing. Thus, $f_x(0, 5) < 0$.

   (b) Moving in the positive y direction near this point the surface slopes up as the function increases, so $f_y(0, 5) > 0$.

13. (a) Estimate $\partial P/\partial r$ and $\partial P/\partial L$ by using difference quotients and reading values of P from the graph:

$$\frac{\partial P}{\partial r}(8, 4000) \approx \frac{P(16, 4000) - P(8, 4000)}{16 - 8}$$
$$= \frac{100 - 80}{8} = 2.5,$$

and

$$\frac{\partial P}{\partial L} \approx \frac{P(8, 5000) - P(8, 4000)}{5000 - 4000}$$
$$= \frac{100 - 80}{1000} = 0.02.$$

$P_r(8, 4000) \approx 2.5$ means that at an interest rate of 8% and a loan amount of \$4000 the monthly payment increases by approximately \$2.50 for every one percent increase of the interest rate. $P_L(8, 4000) \approx 0.02$ means the monthly payment increases by approximately \$0.02 for every \$1 increase in the loan amount at an 8% rate and a loan amount of \$4000.

   (b) Using difference quotients and reading from the graph

$$\frac{\partial P}{\partial r}(8, 6000) \approx \frac{P(14, 6000) - P(8, 6000)}{14 - 8}$$
$$= \frac{140 - 120}{6} = 3.33,$$

and

$$\frac{\partial P}{\partial L}(8, 6000) \approx \frac{P(8, 7000) - P(8, 6000)}{7000 - 6000}$$
$$= \frac{140 - 120}{1000} = 0.02.$$

Again, we see that the monthly payment increases with increases in interest rate and loan amount. The interest rate is $r = 8\%$ as in part (a), but here the loan amount is $L = \$6000$. Since $P_L(8, 4000) \approx P_L(8, 6000)$, the increase in monthly payment per unit increase in loan amount remains the same as in part a). However, in this case, the effect of the interest rate is different: here the monthly payment increases by approximately \$3.33 for every one percent increase of interest rate at $r = 8\%$ and loan amount of \$6000.

   (c)

$$\frac{\partial P}{\partial r}(13, 7000) \approx \frac{P(19, 7000) - P(13, 7000)}{19 - 13}$$
$$= \frac{180 - 160}{6} = 3.33,$$

and

$$\frac{\partial P}{\partial L}(13, 7000) \approx \frac{P(13, 8000) - P(13, 7000)}{8000 - 7000}$$
$$= \frac{180 - 160}{1000} = 0.02.$$

The figures show that the rates of change of the monthly payment with respect to the interest rate and loan amount are roughly the same for $(r, L) = (8, 6000)$ and $(r, L) = (13, 7000)$.

17.

TABLE 13.1 *Estimated values of*
$H(T, w)$ *(in calories/meter^3)*

T	w (gm/m^3)			
(°C)	0.1	0.2	0.3	0.4
10	110	240	330	450
20	100	180	260	350
30	70	150	220	300
40	65	140	200	270

Values of H from the graph are given in Table 13.1. In order to compute $H_w(T, w)$ for $w = 0.3$, it is useful to have the column corresponding to $w = 0.4$. The row corresponding to $T = 40$ is not used in this problem. The partial derivative $H_w(T, w)$ can be approximated by

$$H_w(10, 0.1) \approx \frac{H(10, 0.1 + h) - H(10, 0.1)}{h} \quad \text{for small } h.$$

We choose $h = 0.1$ because we can read off a value for $H(10, 0.2)$ from the graph. If we take $H(10, 0.2) = 240$, we get the approximation

$$H_w(10, 0.1) \approx \frac{H(10, 0.2) - H(10, 0.1)}{0.1} = \frac{240 - 110}{0.1} = 1300.$$

In practical terms, we have found that for fog at $10°$ C containing 0.1 g/m^3 of water, an increase in the water content of the fog will increase the heat requirement for dissipating the fog at the rate given by $H_w(10, 0.1)$. Specifically, a 1 g/m^3 increase in the water content will increase the heat required to dissipate the fog by about 1300 calories per cubic meter of fog.

Wetter fog is harder to dissipate. Other values of $H_w(T, w)$ in Table 13.2 are computed using the formula

$$H_w(T, w) \approx \frac{H(T, w + 0.1) - H(T, w)}{0.1},$$

where we have used Table 13.1 to evaluate H.

TABLE 13.2 *Table of values of*
$H_w(T, w)$ *(in cal/gm)*

T	w (gm/m^3)		
(°C)	0.1	0.2	0.3
10	1300	900	1200
20	800	800	900
30	800	700	800

Solutions for Section 13.2

1. Since $f_y(3, 2)$ equals the derivative of $f(3, y)$ at $y = 2$, we use the function

$$f(3, y) = \frac{9}{y + 1}.$$

Differentiating with respect to y, we get

$$f_y(3, y) = \frac{d}{dy}\left(\frac{9}{y+1}\right) = \frac{-9}{(y+1)^2},$$

and so

$$f_y(3, 2) = -1.$$

5. Differentiating with respect to x gives

$$g_x(x, y) = \frac{\partial}{\partial x}\ln(ye^{xy}) = (ye^{xy})^{-1}\frac{\partial}{\partial x}(ye^{xy}) = (ye^{xy})^{-1} \cdot y\frac{\partial}{\partial x}(e^{xy})$$
$$= (ye^{xy})^{-1} \cdot y \cdot y \cdot e^{xy}$$
$$= y$$

9. $\frac{\partial}{\partial y}(3x^5y^7 - 32x^4y^3 + 5xy) = 21x^5y^6 - 96x^4y^2 + 5x$

13. $\frac{\partial}{\partial B}\left(\frac{1}{u_0}B^2\right) = \frac{2B}{u_0}$

17. $\frac{\partial F}{\partial m_2} = \frac{Gm_1}{r^2}$

21. $u_E = \frac{1}{2}\epsilon_0 \cdot 2E + 0 = \epsilon_0 E$

25. $z_x = \frac{1}{2ay}(-2)\frac{1}{x^3} + \frac{15x^4abc}{y} = -\frac{1}{ax^3y} + \frac{15abcx^4}{y} = \frac{15a^2bcx^7 - 1}{ax^3y}$

29.

$$\frac{\partial}{\partial w}(\sqrt{2\pi xyw - 13x^7y^3v}) = \frac{1}{2}(2\pi xyw - 13x^7y^3v)^{-1/2}(2\pi xy - 0)$$
$$= \frac{\pi xy}{\sqrt{2\pi xyw - 13x^7y^3v}}$$

33. We regard x as constant and differentiate with respect to y using the product rule:

$$\frac{\partial z}{\partial y} = 2e^{x+2y}\sin y + e^{x+2y}\cos y$$

Substituting $x = 1, y = 0.5$ gives

$$\left.\frac{\partial z}{\partial y}\right|_{(1,0.5)} = 2e^2\sin(0.5) + e^2\cos(0.5) = 13.6.$$

37. (a) To calculate $\partial B/\partial t$, we hold P constant and differentiate B with respect to t:

$$\frac{\partial B}{\partial t} = \frac{\partial}{\partial t}(Pe^{rt}) = Pre^{rt}.$$

In financial terms, $\partial B/\partial t$ represents the change in the amount of money in the bank as one unit of time passes by.

(b) To calculate $\partial B/\partial P$, we hold t constant and differentiate B with respect to P:

$$\frac{\partial B}{\partial P} = \frac{\partial}{\partial P}(Pe^{rt}) = e^{rt}.$$

In financial terms, $\partial B/\partial P$ represents the change in the amount of money in the bank at time t as you increase the amount of money that was initially deposited by one unit.

Solutions for Section 13.3

1. We have
$$z = e^y + x + x^2 + 6.$$

The partial derivatives are

$$\left.\frac{\partial z}{\partial x}\right|_{(x,y)=(1,0)} = (2x+1)\Big|_{(x,y)=(1,0)} = 3$$

$$\left.\frac{\partial z}{\partial y}\right|_{(x,y)=(1,0)} = e^y\Big|_{(x,y)=(1,0)} = 1.$$

So the equation of the tangent plane is

$$z = 9 + 3(x-1) + y = 6 + 3x + y.$$

5. We have
$$f_x(3,1) = \left.\frac{\partial f}{\partial x}\right|_{(3,1)} = 2xy|_{(3,1)} = 6,$$

and

$$f_y(3,1) = \left.\frac{\partial f}{\partial y}\right|_{(3,1)} = x^2|_{(3,1)} = 9.$$

Also $f(3,1) = 9$. So the local linearization is,

$$z = 9 + 6(x-3) + 9(y-1).$$

9. (a) The linear approximation gives

$$f(520,24) \approx 24.20, \quad f(480,24) \approx 23.18,$$
$$f(500,22) \approx 25.52, \quad f(500,26) \approx 21.86.$$

The approximations for $f(520,24)$ and $f(500,26)$ agree exactly with the values in the table; the other two do not. The reason for this is that the partial derivatives were estimated using difference quotients with these values.

(b) We could get a more balanced estimate by using a difference quotient that uses the values on both sides. Thus, we could estimate the partial derivatives as follows:

$$f_T(500,24) \approx \frac{f(520,24) - f(480,24)}{40}$$
$$= \frac{(24.20 - 23.19)}{40} = 0.02525,$$

and

$$f_p(500,24) \approx \frac{f(500,26) - f(500,22)}{4}$$
$$= \frac{(21.86 - 25.86)}{4} = -1.$$

This yields the linear approximation

$$V = f(T,p) \approx 23.69 + 0.02525(T - 500) - (p - 24) \text{ ft}^3.$$

This approximation yields values

$$f(520, 24) \approx 24.195, \quad f(480, 24) \approx 23.185,$$
$$f(500, 22) \approx 25.69, \quad f(500, 26) \approx 21.69.$$

Although none of these predictions are accurate, the error in the predictions that were wrong before has been reduced. This new linearization is a better all-round approximation for values near $(500, 24)$.

13.   Since $g_u = 2u + v$ and $g_v = u$, we have

$$dg = (2u + v)\, du + u\, dv$$

17.   We have $dP = P_L dL + P_K dK$.
      Now

$$P_K = (1.01)(0.75)K^{-0.25}L^{0.25}$$

$$P_K(100, 1) \approx 2.395,$$

and

$$P_L = (1.01)(0.25)K^{0.75}L^{-0.75}$$

$$P_L(100, 1) \approx 0.008$$

Thus

$$dP \approx 2.395\, dK + 0.008\, dL$$

21.   (a)   If the volume is held constant, $\Delta V = 0$, so $\Delta U \approx 27.32\Delta T$. Thus the energy increases if the temperature increases.
      (b)   If the temperature is held constant, then $\Delta T = 0$, so $\Delta U \approx 840\Delta V$. Thus the energy increases if the volume increases (yes, it sounds bizarre, but remember the temperature is being held constant).
      (c)   First, we convert 100 cm^3 to 0.0001 m^3. Now, using the differential approximation,

$$\begin{aligned} \Delta U &\approx 840\, \Delta V + 27.32\, \Delta T \\ &= (840)(-0.0001) + (27.32)(2) \\ &= -0.084 + 54.64 \approx 55 \text{ joules.} \end{aligned}$$

25.   (a)   The area of a circle of radius r is given by

$$A = \pi r^2$$

and the perimeter is

$$L = 2\pi r.$$

Thus we get

$$r = \frac{L}{2\pi}$$

and

$$A = \pi \left(\frac{L}{2\pi} \right)^2 = \frac{L^2 \pi}{4\pi^2} = \frac{L^2}{4\pi}.$$

Thus we get

$$\pi = \frac{L^2}{4A}.$$

(b) We will treat π as a function of L and A.

$$d\pi = \frac{\partial \pi}{\partial L} dL + \frac{\partial \pi}{\partial A} dA = \frac{2L}{4A} dL - \frac{L^2}{4A^2} dA.$$

If L is in error by a factor λ, then $\Delta L = \lambda L$, and if A is in error by a factor μ, then $\Delta A = \mu A$. Therefore,

$$\Delta \pi \approx \frac{2L}{4A} \Delta L - \frac{L^2}{4A^2} \Delta A$$

$$= \frac{2L}{4A} \lambda L - \frac{L^2}{4A^2} \mu A$$

$$= \frac{2\lambda L^2}{4A} - \frac{\mu L^2}{4A} = (2\lambda - \mu)\frac{L^2}{4A} = (2\lambda - \mu)\pi,$$

so π is in error by a factor of $2\lambda - \mu$.

Solutions for Section 13.4

1. (a) We use the definition of the directional derivative

$$f_{\vec{u}}(a, b) = \lim_{h \to 0} \frac{f(a + hu_1, b + hu_2) - f(a, b)}{h}$$

where $\vec{u} = u_1 \vec{i} + u_2 \vec{j}$ is a unit vector. In this case the vector $\vec{u}$ is the direction of the point $(3, 5)$ and so is parallel to $(3 - 1)\vec{i} + (5 - 4)\vec{j}$. Thus the unit vector

$$\vec{u} = \frac{2}{\sqrt{5}}\vec{i} + \frac{1}{\sqrt{5}}\vec{j} \approx 0.894\vec{i} + 0.447\vec{j}.$$

We approximate by taking a small value of h, giving

$$f_{\vec{u}}(1, 4) \approx \frac{f(1 + 0.894h, 4 + 0.447h) - f(1, 4)}{h}$$

We choose a value of h say $h = 0.01$

$$f_{\vec{u}}(1, 4) \approx \frac{f(1.00894, 4.00447) - f(1, 4)}{0.01}$$

$$= \frac{1.00984 + \ln(4.00447) - 1 - \ln 4}{0.01}$$

$$= 1.01$$

(b) Here $\vec{u}$ is the same as in part (a), and we take $h = 0.01$, giving

$$f_{\vec{u}}(3, 5) \approx \frac{f(3, 5) - f(3 - 0.894h, 5 - 0.447h)}{h}$$

$$= \frac{f(3, 5) - f(2.99106, 4.99553)}{h}$$

$$= \frac{3 + \ln 5 - 2.99106 - \ln(4.99553)}{0.01}$$

$$= 0.98$$

5. Since $\vec{u} = (\vec{i} - \vec{j})/\sqrt{2}$, we head away from the point $(3, 1)$ toward the point $(4, 0)$.
    From the graph, we see that $f(3, 1) = 1$ and $f(4, 0) = 4$. Since the points $(3, 1)$ and $(4, 0)$ are a distance $\sqrt{2}$
    apart, we have

$$f_{\vec{u}}(3, 1) \approx \frac{f(4, 0) - f(3, 1)}{\sqrt{2}} = \frac{4 - 1}{\sqrt{2}} = 2.12.$$

9. (a) In the $\vec{i} - \vec{j}$ direction the function is decreasing, so the value of $g_{\vec{u}}(2, 5)$ is negative.
    (b) In the $\vec{i} + \vec{j}$ direction the function is decreasing, so the value of $g_{\vec{u}}(2, 5)$ negative as well.

13. Since the partial derivatives are

$$z_x = e^y, \quad \text{and} z_y = xe^y + e^y + ye^y,$$

we have

$$\nabla z = e^y \vec{i} + e^y (1 + x + y) \vec{j}.$$

17. Since the partial derivatives are

$$\frac{\partial f}{\partial m} = 2m + 0 = 2m$$
$$\frac{\partial f}{\partial n} = 0 + 2n = 2n$$

we have

$$\text{grad } f = \frac{\partial f}{\partial m}\vec{i} + \frac{\partial f}{\partial n}\vec{j} = 2m\vec{i} + 2n\vec{j}.$$

21. Since the partial derivatives are

$$\frac{\partial f}{\partial \alpha} = \frac{1}{2}(5\alpha^2 + \beta)^{-1/2}(10\alpha + 0) = \frac{5\alpha}{\sqrt{5\alpha^2 + \beta}}$$
$$\frac{\partial f}{\partial \beta} = \frac{1}{2}(5\alpha^2 + \beta)^{-1/2}(0 + 1) = \frac{1}{2\sqrt{5\alpha^2 + \beta}},$$

we have

$$\text{grad } f = \frac{\partial f}{\partial \alpha}\vec{i} + \frac{\partial f}{\partial \beta}\vec{j} = \left(\frac{5\alpha}{\sqrt{5\alpha^2 + \beta}}\right)\vec{i} + \left(\frac{1}{2\sqrt{5\alpha^2 + \beta}}\right)\vec{j}.$$

25. Since the partial derivatives are

$$\frac{\partial f}{\partial x} = \frac{1}{2}(\tan x + y)^{-1/2}\left(\frac{1}{\cos^2 x} + 0\right) = \frac{1}{2\cos^2 x\sqrt{\tan x + y}},$$

and

$$\frac{\partial f}{\partial y} = \frac{1}{2}(\tan x + y)^{-1/2}(0 + 1) = \frac{1}{2\sqrt{\tan x + y}},$$

then

$$\text{grad } f = \frac{\partial f}{\partial x}\vec{i} + \frac{\partial f}{\partial y}\vec{j} = \left(\frac{1}{2\cos^2 x\sqrt{\tan x + y}}\right)\vec{i} + \left(\frac{1}{2\sqrt{\tan x + y}}\right)\vec{j}.$$

Hence we have

$$\text{grad } f\bigg|_{(0,1)} = \left(\frac{1}{2(\cos(0))^2 \sqrt{\tan(0)+1}}\right)\vec{i} + \left(\frac{1}{2\sqrt{\tan(0)+1}}\right)\vec{j}$$

$$= \left(\frac{1}{2(1)^2\sqrt{0+1}}\right)\vec{i} + \left(\frac{1}{2\sqrt{0+1}}\right)\vec{j}$$

$$= \frac{1}{2}\vec{i} + \frac{1}{2}\vec{j}.$$

29. (a) The partial derivatives are given by

$$f_x = e^x(\tan y) + 4xy, \quad f_y = e^x(\sec^2 y) + 2x^2.$$

Thus

$$f_x(0, \frac{\pi}{4}) = 1 \quad \text{and} \quad f_y(0, \frac{\pi}{4}) = 2,$$

and so

$$\text{grad } f(0, \frac{\pi}{4}) = \vec{i} + 2\vec{j}.$$

The unit vector $\vec{u_1}$ in the direction of $\vec{i} - \vec{j}$ is $\frac{1}{\sqrt{2}}(\vec{i} - \vec{j})$. Then the directional derivative of f at $(0, \frac{\pi}{4})$ in the direction of $\vec{i} - \vec{j}$ is

$$f_{\vec{u_1}}(0, \frac{\pi}{4}) = \text{grad } f(0, \frac{\pi}{4}) \cdot \vec{u_1}$$

$$= (\vec{i} + 2\vec{j}) \cdot (\frac{1}{\sqrt{2}}\vec{i} - \frac{1}{\sqrt{2}}\vec{j})$$

$$= \frac{1}{\sqrt{2}} - \sqrt{2} = -\frac{\sqrt{2}}{2}.$$

(b) The unit vector $\vec{u_2}$ in the direction of $\vec{i} + \sqrt{3}\vec{j}$ is $\vec{u_2} = \frac{1}{2}(\vec{i} + \sqrt{3}\vec{j})$. From part (a),

$$\text{grad } f(0, \frac{\pi}{4}) = \vec{i} + 2\vec{j}.$$

Then the directional derivative of f at $(0, \frac{\pi}{4})$ in the direction of $\vec{i} + \sqrt{3}\vec{j}$ is

$$f_{\vec{u_2}}(0, \frac{\pi}{4}) = \text{grad } f(0, \frac{\pi}{4}) \cdot \vec{u_2} = (\vec{i} + 2\vec{j}) \cdot (\frac{1}{2}\vec{i} + \frac{\sqrt{3}}{2}\vec{j})$$

$$= \frac{1}{2} + \sqrt{3}.$$

33. Directional derivative $= \nabla f \cdot \vec{u}$, where $\vec{u} =$ unit vector. If we move from $(4, 5)$ to $(5, 6)$, we move in the direction $\vec{i} + \vec{j}$ so $\vec{u} = \frac{1}{\sqrt{2}}\vec{i} + \frac{1}{\sqrt{2}}\vec{j}$. So,

$$\nabla f \cdot \vec{u} = f_x\left(\frac{1}{\sqrt{2}}\right) + f_y\left(\frac{1}{\sqrt{2}}\right) = 2.$$

Similarly, if we move from $(4, 5)$ to $(6, 6)$, the direction is $2\vec{i} + \vec{j}$ so $\vec{u} = \frac{2}{\sqrt{5}}\vec{i} + \frac{1}{\sqrt{5}}\vec{j}$. So

$$\nabla f \cdot \vec{u} = f_x\left(\frac{2}{\sqrt{5}}\right) + f_y\left(\frac{1}{\sqrt{5}}\right) = 3.$$

Solving the system of equations for f_x and f_y

$$f_x + f_y = 2\sqrt{2}$$

$$2f_x + f_y = 3\sqrt{5}$$

gives

$$f_x = 3\sqrt{5} - 2\sqrt{2}$$
$$f_y = 4\sqrt{2} - 3\sqrt{5}.$$

Thus at $(4, 5)$,

$$\nabla f = (3\sqrt{5} - 2\sqrt{2})\vec{i} + (4\sqrt{2} - 3\sqrt{5})\vec{j}.$$

37. $\|\nabla f\|$ at P is larger because the level curves are closer there.

Solutions for Section 13.5

1.   The unit vector $\vec{u}$ in the direction of $\vec{v} = 2\vec{i} + \vec{j} - 2\vec{k}$ is

$$\vec{u} = \frac{\vec{v}}{\|\vec{v}\|} = \left(\frac{2}{3}\right)\vec{i} + \left(\frac{1}{3}\right)\vec{j} - \left(\frac{2}{3}\right)\vec{k}.$$

The partial derivatives are

$$f_x(x, y, z) = 2x + 3y,$$
$$f_y(x, y, z) = 3x,$$
$$f_z(x, y, z) = 2.$$

Hence,

$$f_{\vec{u}}(2, 0, -1) = f_x(2, 0, -1)\left(\frac{2}{3}\right) + f_y(2, 0, -1)\left(\frac{1}{3}\right) + f_z(2, 0, -1)\left(-\frac{2}{3}\right)$$

$$= 4\left(\frac{2}{3}\right) + 6\left(\frac{1}{3}\right) + 2\left(-\frac{2}{3}\right)$$

$$= \frac{10}{3}$$

5.   The surface is given by $F(x, y, z) = 0$ where $F(x, y, z) = x - y^3 z^7$. The normal direction is $\nabla F = \frac{\partial F}{\partial x}\vec{i} + \frac{\partial F}{\partial y}\vec{j} + \frac{\partial F}{\partial z}\vec{k} = \vec{i} - 3y^2 z^7\vec{j} - 7y^3 z^6\vec{k}$.

Thus, at $(1, -1, -1)$ a normal vector is $\vec{i} + 3\vec{j} + 7\vec{k}$. The tangent plane has the equation

$$\left[(x\vec{i} + y\vec{j} + z\vec{k}) - (\vec{i} - \vec{j} - \vec{k})\right] \cdot (\vec{i} + 3\vec{j} + 7\vec{k}) = 0, \text{ that is,}$$

$$x + 3y + 7z = -9.$$

9. (a)

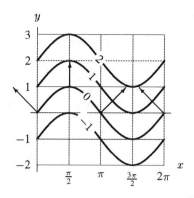

Figure 13.1

(b) The bug is walking parallel to the y-axis. Looking to the right or left, the bug sees higher contours — thus it is in a valley.

(c) See Figure 13.1.

13. The point $(4, 1, 3)$ lies on the surface. The surface is the level surface of the function

$$F(x, y, z) = f(x, y) - z = 0.$$

The normal to the surface at the point $(4, 1, 3)$ is

$$\text{grad } F(4, 1, 3) = f_x(4, 1)\vec{i} + f_y(4, 1)\vec{j} - \vec{k} = 2\vec{i} - \vec{j} - \vec{k}.$$

Thus the equation of the tangent plane is

$$2x - y - z = 2(4) - 1 - 3 = 4$$
$$2x - y - z = 4$$

17. If write $\vec{r} = x\vec{i} + y\vec{j} + z\vec{k}$, then we know

$$\text{grad } f(x, y, z) = g(x, y, z)(x\vec{i} + y\vec{j} + z\vec{k}) = g(x, y, z)\vec{r}$$

so grad f is everywhere radially outward, and therefore perpendicular to a sphere centered at the origin. If f were not constant on such a sphere, then grad f would have a component tangent to the sphere. Thus, f must be constant on any sphere centered at the origin.

Solutions for Section 13.6

1. Using the chain rule we see:

$$
\begin{aligned}
\frac{dz}{dt} &= \frac{\partial z}{\partial x}\frac{dx}{dt} + \frac{\partial z}{\partial y}\frac{dy}{dt} \\
&= -y^2 e^{-t} + 2xy \cos t \\
&= -(\sin t)^2 e^{-t} + 2e^{-t} \sin t \cos t \\
&= \sin(t)e^{-t}(2 \cos t - \sin t)
\end{aligned}
$$

We can also solve the problem using one variable methods:

$$z = e^{-t}(\sin t)^2$$
$$\frac{dz}{dt} = \frac{d}{dt}(e^{-t}(\sin t)^2)$$
$$= \frac{de^{-t}}{dt}(\sin t)^2 + e^{-t}\frac{d(\sin t)^2}{dt}$$
$$= -e^{-t}(\sin t)^2 + 2e^{-t}\sin t \cos t$$
$$= e^{-t}\sin t(2\cos t - \sin t)$$

5. Substituting into the chain rule gives

$$\frac{dz}{dt} = \frac{\partial z}{\partial x}\frac{dx}{dt} + \frac{\partial z}{\partial y}\frac{dy}{dt} = e^y(2) + xe^y(-2t)$$
$$= 2e^y(1 - xt) = 2e^{1-t^2}(1 - 2t^2).$$

9. Since z is a function of two variables x and y which are functions of two variables u and v, the two chain rule identities which apply are:

$$\frac{\partial z}{\partial u} = \frac{\partial z}{\partial x}\frac{\partial x}{\partial u} + \frac{\partial z}{\partial y}\frac{\partial y}{\partial u} = e^y(\frac{1}{u}) + xe^y \cdot 0 = \frac{e^v}{u}.$$
$$\frac{\partial z}{\partial v} = \frac{\partial z}{\partial x}\frac{\partial x}{\partial v} + \frac{\partial z}{\partial y}\frac{\partial y}{\partial v} = e^y(0) + xe^y \cdot 1 = e^v \ln u.$$

13. Since z is a function of two variables x and y which are functions of two variables u and v, the two chain rule identities which apply are:

$$\frac{\partial z}{\partial u} = \frac{\partial z}{\partial x}\frac{\partial x}{\partial u} + \frac{\partial z}{\partial y}\frac{\partial y}{\partial u} = \left(\cos\left(\frac{x}{y}\right)\right)\left(\frac{1}{y}\right)\frac{1}{u} + \left(\cos\left(\frac{x}{y}\right)\right)\left(\frac{-x}{y^2}\right)\cdot 0$$
$$= \frac{1}{vu}\cos\left(\frac{\ln u}{v}\right).$$
$$\frac{\partial z}{\partial v} = \frac{\partial z}{\partial x}\frac{\partial x}{\partial v} + \frac{\partial z}{\partial y}\frac{\partial y}{\partial v} = \left(\cos\left(\frac{x}{y}\right)\right)\left(\frac{1}{y}\right)\cdot 0 + \left(\cos\left(\frac{x}{y}\right)\right)\left(\frac{-x}{y^2}\right)\cdot 1 = -\frac{\ln u}{v^2}\cos\left(\frac{\ln u}{v}\right).$$

17.

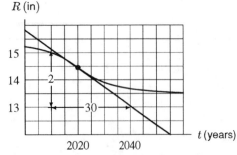

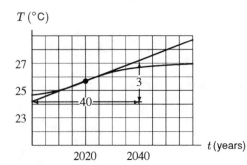

Figure 13.2: Global warming predictions: Rainfall as a function of time

Figure 13.3: Global warming predictions: Temperature as a function of time

We know that, as long as the temperature and rainfall stay close to their current values of $R = 15$ inches and $T = 30°C$, a change, ΔR, in rainfall and a change, ΔT, in temperature produces a change, ΔC, in corn production given by

$$\Delta C \approx 3.3\Delta R - 5\Delta T.$$

Now both R and T are functions of time t (in years), and we want to find the effect of a small change in time, Δt, on R and T. Figure 13.2 shows that the slope of the graph for R versus t is about $-2/30 \approx -0.07$ in/year when $t = 2020$. Similarly, Figure 13.3 shows the slope of the graph of T versus t is about $3/40 \approx 0.08°C$/year when $t = 2020$. Thus, around the year 2020,

$$\Delta R \approx -0.07\Delta t \quad \text{and} \quad \Delta T \approx 0.08\Delta t.$$

Substituting these into the equation for ΔC, we get

$$\Delta C \approx (3.3)(-0.07)\Delta t - (5)(0.08)\Delta t \approx -0.6\Delta t.$$

Since at present $C = 100$, corn production will decline by about 0.6 % between the years 2020 and 2021. Now $\Delta C \approx -0.6\Delta t$ tells us that when $t = 2020$,

$$\frac{\Delta C}{\Delta t} \approx -0.6, \quad \text{and therefore, that} \quad \frac{dC}{dt} \approx -0.6.$$

21. We will use analysis similar to that in Example 6. Since V is a function of P and T, we have

$$dV = \left(\frac{\partial V}{\partial T}\right)_P dT + \left(\frac{\partial V}{\partial P}\right)_T dP$$

We are interested in $\left(\frac{\partial U}{\partial V}\right)_T$ so we use the formula for dU corresponding to U_2. Substituting g for dV into this formula for dU gives

$$dU = \left(\frac{\partial U}{\partial T}\right)_V dT + \left(\frac{\partial U}{\partial V}\right)_T \left(\left(\frac{\partial V}{\partial T}\right)_P dT + \left(\frac{\partial V}{\partial P}\right)_T dP\right)$$

$$= \left(\left(\frac{\partial U}{\partial T}\right)_V + \left(\frac{\partial U}{\partial V}\right)_T \left(\frac{\partial V}{\partial T}\right)_P\right) dT + \left(\frac{\partial U}{\partial V}\right)_T \left(\frac{\partial V}{\partial P}\right)_T dP$$

But we are also interested in $\left(\frac{\partial U}{\partial P}\right)_T$ so we compare with the formula for dU corresponding to U_1.

$$dU = \left(\frac{\partial U}{\partial T}\right)_P dT + \left(\frac{\partial U}{\partial P}\right)_T dP.$$

Since the coefficients of dP must be identical, we get

$$\left(\frac{\partial U}{\partial P}\right)_T = \left(\frac{\partial U}{\partial V}\right)_T \left(\frac{\partial V}{\partial P}\right)_T.$$

25. Use chain rule for the equation $0 = F(x, y, f(x, y))$. Differentiating both sides with respect to x, remembering $z = f(x, y)$ and regarding y as a constant gives:

$$0 = \frac{\partial F}{\partial x}\frac{dx}{dx} + \frac{\partial F}{\partial z}\frac{dz}{dx}.$$

Since $dx/dx = 1$, we get

$$-\frac{\partial F}{\partial x} = \frac{\partial F}{\partial z}\frac{\partial z}{\partial x},$$

so

$$\frac{\partial z}{\partial x} = -\frac{\partial F/\partial x}{\partial F/\partial z}.$$

Similarly, differentiating both sides of the equation $0 = F(x, y, f(x, y))$ with respect to y gives:

$$0 = \frac{\partial F}{\partial y}\frac{dy}{dy} + \frac{\partial F}{\partial z}\frac{dz}{dy}.$$

Since $dy/dy = 1$, we get

$$-\frac{\partial F}{\partial y} = \frac{\partial F}{\partial z}\frac{\partial z}{\partial y},$$

so

$$\frac{\partial z}{\partial y} = -\frac{\partial F/\partial y}{\partial F/\partial z}.$$

Solutions for Section 13.7

1. Calculating the partial derivatives:

$$\frac{\partial f}{\partial x} = 2(x + y), \qquad \frac{\partial^2 f}{\partial x^2} = 2.$$

Therefore, we get

$$\frac{\partial f}{\partial y} = 2(x + y), \qquad \frac{\partial^2 f}{\partial y^2} = 2, \qquad \frac{\partial^2 f}{\partial y \partial x} = 2, \qquad \frac{\partial^2 f}{\partial x \partial y} = 2.$$

5. Since $f(x, y) = \sin(x^2 + y^2)$, we have

$$f_x = (\cos(x^2 + y^2))2x \quad , f_y = (\cos(x^2 + y^2))2y$$
$$f_{xx} = -(\sin(x^2 + y^2))4x^2 + 2\cos(x^2 + y^2)$$
$$f_{xy} = -(\sin(x^2 + y^2))4xy = f_{yx}$$
$$f_{yy} = -(\sin(x^2 + y^2))4y^2 + 2\cos(x^2 + y^2).$$

9. Since $z_y = g(x), z_{yy} = 0$, because g is a function of x only.

13. (a) $f_x(P) < 0$ because f decreases as you go to the right.
    (b) $f_y(P) = 0$ because f does not change as you go up.
    (c) $f_{xx}(P) < 0$ because f_x decreases as you go to the right (f_x changes from a small negative number to a large negative number).
    (d) $f_{yy}(P) = 0$ because f_y does not change as you go up.
    (e) $f_{xy}(P) = 0$ because f_x does not change as you go up.

17. (a) $f_x(P) < 0$ because f decreases as you go to the right.
    (b) $f_y(P) < 0$ because f decreases as you go up.
    (c) $f_{xx}(P) = 0$ because f_x does not change as you go to the right. (Notice that the level curves are equidistant and parallel, so the partial derivatives of f do not change if you move horizontally or vertically.)
    (d) $f_{yy}(P) = 0$ because f_y does not change as you go up.
    (e) $f_{xy}(P) = 0$ because f_x does not change as you go up.

Solutions for Section 13.8

1. (a) We must first find the estimated rate of change from $t = 0$ to $t = 1$, with $u_t(4,0)$ and $u_t(8,0)$. From the heat equation, we know that we must first approximate $u_{xx}(4,0)$ and $u_{xx}(8,0)$. As in example 3, we will use the fact that u_{xx} is approximately a difference quotient of u_x, with the estimated slopes of two nearby points.

    First for $u(4, 1)$:

    $$u_x(3,0) \approx \frac{(u(4,0) - u(2,0))}{(4 - 2)} = \frac{(56 - 52)}{2} = 2$$

    $$u_x(5,0) \approx \frac{(u(6,0) - u(4,0))}{(6 - 4)} = \frac{(62 - 56)}{2} = 3$$

    $$u_{xx}(4,0) \approx \frac{(u_x(5,0) - u_x(3,0))}{(5 - 3)} = \frac{(3 - 2)}{2} = 0.5$$

    $$u_t(4,0) = 0.1u_{xx}(4,0) \approx 0.1(0.5) = 0.05$$

    Now for $u(8, 1)$:

    $$u_x(7,0) \approx \frac{(u(8,0) - u(6,0))}{(8 - 6)} = \frac{(70 - 62)}{2} = 4$$

    $$u_x(9,0) \approx \frac{(u(10,0) - u(8,0))}{(10 - 8)} = \frac{(80 - 70)}{2} = 5$$

    $$u_{xx}(8,0) \approx \frac{(u_x(9,0) - u_x(7,0))}{(9 - 7)} = \frac{(5 - 4)}{2} = 0.5$$

    $$u_t(8,0) = 0.1u_{xx}(8,0) \approx 0.1(0.5) = 0.05$$

    So, using the local linear approximation we have:

    $$u(4,1) \approx u(4,0) + u_t(4,0)(1) \approx 56 + 0.05(1) = 56.05°C$$
    $$u(8,1) \approx u(8,0) + u_t(8,0)(1) \approx 70 + 0.05(1) = 70.05°C.$$

    (b) Now, we will follow the same process for $u(6, 2)$.

    $$u_x(5,1) \approx \frac{(u(6,1) - u(4,1))}{(6 - 4)} = \frac{(62.05 - 56.05)}{2} = 3$$

    $$u_x(7,1) \approx \frac{(u(8,1) - u(6,1))}{(8 - 6)} = \frac{(70.05 - 62.05)}{2} = 4$$

    $$u_{xx}(6,1) \approx \frac{(u_x(7,1) - u_x(5,1))}{(7 - 5)} = \frac{(4 - 3)}{2} = 0.5$$

    $$u_t(6,1) = 0.1u_{xx}(6,1) \approx 0.1(0.5) = 0.05$$

    So, we have

    $$u(6,2) \approx u(6,1) + u_t(6,1)(1) \approx 62.05 + (.05)(1) = 62.1°C.$$

5. Differentiating, we get

$$F_x = -e^{-x} \sin y, \; F_y = e^{-x} \cos y, \; F_{xx} = e^{-x} \sin y, \; F_{yy} = -e^{-x} \sin y = -F_{xx}.$$

Thus, $F_{xx} + F_{yy} = 0$.

9. Write $V = f(u)$ where $u = x + ct$, then using the chain rule

$$\frac{\partial V}{\partial x} = \frac{df}{du} \cdot \frac{\partial u}{\partial x} = f'(u)(1).$$

Similarly,

$$\frac{\partial V}{\partial t} = \frac{df}{du} \cdot \frac{\partial u}{\partial t} = f'(u)(c) = cf'(u).$$

Thus

$$\frac{\partial V}{\partial t} = cf'(u) = c\frac{\partial V}{\partial x}.$$

13. First let us take the partials.

$$u_t = ae^{at} \sin{(bx)}$$
$$u_x = be^{at} \cos{(bx)}$$
$$u_{xx} = -b^2 e^{at} \sin{(bx)}$$

Now, plugging them in the equation, we have

$$ae^{at} \sin{(bx)} = u_t = u_{xx} = -b^2 e^{at} \sin{(bx)}$$

So, we have $a = -b^2$.

17. We must first find the partial derivatives:

$$u_x = be^{-at} \cos{(bx)} \sin{(cy)}$$
$$u_{xx} = -b^2 e^{-at} \sin{(bx)} \sin{(cy)}$$
$$u_y = ce^{-at} \sin{(bx)} \cos{(cy)}$$
$$u_{yy} = -c^2 e^{-at} \sin{(bx)} \sin{(cy)}$$
$$u_t = -ae^{-at} \sin{(bx)} \sin{(cy)}$$

Now, substituting them in to the two-dimensional heat equation we have

$$-ae^{-at} \sin{(bx)} \sin{(cy)} = A(-b^2 e^{-at} \sin{(bx)} \sin{(cy)} - c^2 e^{-at} \sin{(bx)} \sin{(cy)})$$
$$= -A(b^2 + c^2)e^{-at} \sin{(bx)} \sin{(cy)}$$

Since this relationship holds for all values of x and y, we must have

$$A = \frac{a}{(b^2 + c^2)}.$$

Notice that $a/(b^2 + c^2)$ is a constant. Since $A > 0$, we must have $a > 0$.

Solutions for Section 13.9

1. The quadratic Taylor expansion about $(0,0)$ is given by

    $$f(x,y) \approx Q(x,y) = f(0,0) + f_x(0,0)x + f_y(0,0)y + \frac{1}{2}f_{xx}(0,0)x^2 + f_{xy}(0,0)xy + \frac{1}{2}f_{yy}(0,0)y^2.$$

    First we find all the relevant derivatives

    $$\begin{aligned}
    f(x,y) &= e^{-2x^2-y^2} \\
    f_x(x,y) &= -4xe^{-2x^2-y^2} \\
    f_y(x,y) &= -2ye^{-2x^2-y^2} \\
    f_{xx}(x,y) &= -4e^{-2x^2-y^2} + 16x^2e^{-2x^2-y^2} \\
    f_{yy}(x,y) &= -2e^{-2x^2-y^2} + 4y^2e^{-2x^2-y^2} \\
    f_{xy}(x,y) &= 8xye^{-2x^2-y^2}
    \end{aligned}$$

    Now we evaluate each of these derivatives at $(0,0)$ and substitute into the formula to get as our final answer:

    $$Q(x,y) = 1 - 2x^2 - y^2$$

5. We have $z(1,1) = 2e$ and the relevant derivatives are:

    $$\begin{aligned}
    z_x(x,y) &= e^y \quad \text{so} \quad z_x(1,1) = e \\
    z_y(x,y) &= e^y(x+1+y) \quad \text{so} \quad z_y(1,1) = 3e \\
    z_{xx}(x,y) &= 0 \quad \text{so} \quad z_{xx}(1,1) = 0 \\
    z_{xy}(x,y) &= e^y \quad \text{so} \quad z_{xy}(1,1) = e \\
    z_{yy}(x,y) &= e^y(x+2+y) \quad \text{so} \quad z_{yy}(1,1) = 4e .
    \end{aligned}$$

    Thus the linear approximation, $L(x,y)$ to $z(x,y)$ at $(1,1)$, is given by:

    $$\begin{aligned}
    z(x,y) \approx L(x,y) &= z(1,1) + z_x(1,1)(x-1) + z_y(1,1)(y-1) \\
    &= 2e + e(x-1) + 3e(y-1) .
    \end{aligned}$$

    The quadratic approximation, $Q(x,y)$ to $z(x,y)$ at $(1,1)$, is given by:

    $$\begin{aligned}
    z(x,y) \approx Q(x,y) &= z(1,1) + z_x(1,1)(x-1) + z_y(1,1)(y-1) \\
    &\quad + \frac{1}{2}z_{xx}(1,1)(x-1)^2 + z_{xy}(1,1)(x-1)(y-1) + \frac{1}{2} + z_{yy}(1,1)(y-1)^2 \\
    &= 2e + e(x-1) + 3e(y-1) + e(x-1)(y-1) + 2e(y-1)^2 .
    \end{aligned}$$

    Now

    $$L(1.1,1.1) = 6.524, \quad Q(1.1,1.1) = 6.605, \quad z(1.1,1.1) = 6.609 .$$

9. The partial derivatives of $z = \dfrac{xe^y}{x + y}$ are:

$$z_x = \frac{e^y(x + y) - xe^y}{(x + y)^2} = \frac{ye^y}{(x + y)^2}$$

$$z_y = \frac{xe^y(x + y) - xe^y}{(x + y)^2} = \frac{e^y(x^2 + xy - x)}{(x + y)^2}$$

$$z_{xx} = -\frac{2ye^y}{(x + y)^3}$$

$$z_{xy} = \frac{e^y(y + 1)(x + y)^2 - ye^y 2(x + y)}{(x + y)^4} = \frac{e^y[(y + 1)(x + y) - 2y]}{(x + y)^3}$$

$$= \frac{e^y(y^2 + xy + x - y)}{(x + y)^3} = z_{yx}$$

$$z_{yy} = \frac{e^y(x^2 + xy - x + x)(x + y)^2 - e^y(x^2 + xy - x)2(x + y)}{(x + y)^4}$$

$$= \frac{e^y[(x^2 + xy)(x + y) - 2(x^2 + xy - x)]}{(x + y)^3}$$

$$= \frac{e^y(x^3 + 2x^2 y + xy^2 - 2x^2 - 2xy + 2x)}{(x + y)^3}$$

The values of the partial derivatives at the point $(1, 1)$ are:

$$z(1, 1) = \frac{e}{2}$$

$$z_x(1, 1) = \frac{e}{4} = z_y(1, 1)$$

$$z_{xx}(1, 1) = \frac{-2}{8}e = \frac{-e}{4}$$

$$z_{xy}(1, 1) = \frac{2}{8}e = \frac{e}{4}$$

$$z_{yy}(1, 1) = \frac{2}{8}e = \frac{e}{4}.$$

Thus the linear approximation, $L(x, y)$ to $z(x, y)$, is given by:

$$z(x, y) \approx L(x, y) = z(1, 1) + z_x(1, 1)(x - 1) + z_y(1, 1)(y - 1)$$
$$= \frac{e}{2} + \frac{e}{4}(x - 1) + \frac{e}{4}(y - 1).$$

The quadratic approximation, $Q(x, y)$ to $z(x, y)$ at $(1, 1)$, is given by:

$$z(x, y) \approx Q(x, y) = z(1, 1) + z_x(1, 1)(x - 1) + z_y(1, 1)(y - 1)$$
$$+ \frac{1}{2}z_{xx}(1, 1)(x - 1)^2 + z_{xy}(1, 1)(x - 1)(y - 1) + \frac{1}{2}z_{yy}(1, 1)(y - 1)^2$$
$$= \frac{e}{2} + \frac{e}{4}(x - 1) + \frac{e}{4}(y - 1) - \frac{e}{8}(x - 1)^2 + \frac{e}{4}(x - 1)(y - 1) + \frac{e}{8}(y - 1)^2.$$

We have:

$$L(1.1, 1.1) = 1.495, \quad Q(1.1, 1.1) = 1.502, \quad z(1.1, 1.1) = 1.502\,.$$

13. The contour diagrams in Figures 13.4–13.9 use the fact that

$$f(x, y) = \sqrt{x + 2y + 1},$$

$$L(x, y) = 1 + \frac{1}{2}x + y,$$

$$Q(x, y) = 1 + \frac{1}{2}x + y - \frac{1}{8}x^2 - \frac{1}{2}xy - \frac{1}{2}y^2.$$

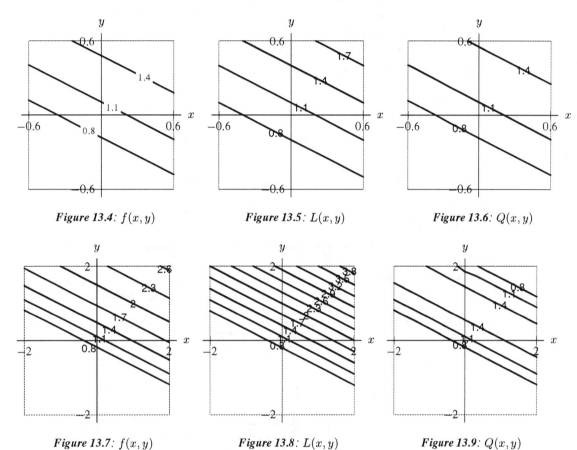

Figure 13.4: $f(x, y)$ Figure 13.5: $L(x, y)$ Figure 13.6: $Q(x, y)$

Figure 13.7: $f(x, y)$ Figure 13.8: $L(x, y)$ Figure 13.9: $Q(x, y)$

The contours for $f(x, y)$ and $L(x, y)$ are straight lines; those for $L(x, y)$ are equally spaced because $L(x, y)$ is a linear function. The contours for $f(x, y)$ are straight lines because if we set

$$f(x, y) = \sqrt{x + 2y + 1} = \text{constant}$$

then

$$x + 2y + 1 = \text{constant}.$$

However, the contours of $f(x, y)$ are not equally spaced because $f(x, y)$ is not linear.

In the "close up" diagram $[-0.6, 0.6] \times [-0.6, 0.6]$, the contours of $Q(x, y)$ look like lines (though they aren't). The contour diagram of $Q(x, y)$ is more similar to the contour diagram of $f(x, y)$ than is $L(x, y)$. This is because $Q(x, y)$ is a better approximation to $f(x, y)$ than is $L(x, y)$.

In the $[-2, 2] \times [-2, 2]$ diagram, the values on the level curves of $L(x, y)$ and $Q(x, y)$ show that neither of them is a good approximation to $f(x, y)$ away from the origin.

17. We have $f(0,0) = 1$ and the relevant derivatives are:

$$\begin{aligned}
f_x(x,y) &= (e^x - 1)\cos y \quad \text{so} \quad f_x(0,0) = 0 \\
f_y(x,y) &= -(e^x - x)\sin y \quad \text{so} \quad f_y(0,0) = 0 \\
f_{xx}(x,y) &= e^x \cos y \quad \text{so} \quad f_{xx}(0,0) = 1 \\
f_{xy}(x,y) &= -(e^x - 1)\sin y \quad \text{so} \quad f_{xy}(0,0) = 0 \\
f_{yy}(x,y) &= -(e^x - x)\cos y \quad \text{so} \quad f_{yy}(0,0) = -1 \\
f_{xxx}(x,y) &= e^x \cos y \\
f_{xxy}(x,y) &= -e^x \sin y \\
f_{xyy}(x,y) &= -(e^x - 1)\cos y \\
f_{yyy}(x,y) &= (e^x - x)\sin y \; .
\end{aligned}$$

(a) Thus,

$$L(x,y) = f(0,0) + f_x(0,0)\cdot x + f_y(0,0)\cdot y = 1$$

and

$$|E_L(x,y)| = |f(x,y) - L(x,y)| \le 2\cdot M_L(x^2 + y^2) \le 0.04 M_L$$

for $|x| \le 0.1$ and $|y| \le 0.1$, where $M_L \ge |f_{xx}|, |f_{xy}|, |f_{yy}|$ for (x,y) such that $d(x,y) = (x^2 + y^2)^{\frac{1}{2}} \le d_0 = (0.1^2 + 0.1^2)^{\frac{1}{2}} = 0.14$.

As $f_{xx}(x,y) = e^x \cos y$, the maximum of $|f_{xx}(x,y)|$ on the disk $d(x,y) \le 0.14$ is $e^{0.14}\cdot\cos 0 \le 1.151$.

Similarly, $f_{xy}(x,y) = -(e^x - 1)\sin y$, and $|f_{xy}(x,y)| \le |\sin y| \le 1$.

Also, $f_{yy}(x,y) = -(e^x - x)\cos y$ so $|f_{yy}(x,y)| \le |e^x - x| \le 1.011$. So we take $M_L = 1.151$ and get

$$|E_L(x,y)| \le 0.047 \quad \text{for} \quad |x| \le 0.1 \quad \text{and} \quad |y| \le 0.1.$$

(b)

$$Q(x,y) = f(0,0) + f_x(0,0)\cdot x + f_y(0,0)\cdot y + \frac{1}{2}f_{xx}(0,0)\cdot x^2$$

$$+ f_{xy}(0,0)\cdot xy + \frac{1}{2}f_{yy}(0,0)\cdot y^2 = 1 + \frac{x^2}{2} - \frac{y^2}{2}.$$

and

$$|E_Q(x,y)| = |f(x,y) - Q(x,y)| \le \frac{4}{3}M_Q(x^2 + y^2)^{\frac{3}{2}} \le \frac{4}{3}M_Q(0.02)^{\frac{3}{2}} = 0.004 M_Q$$

where $M_Q \ge |f_{xxx}|, |f_{xxy}|, |f_{xyy}|, |f_{yyy}|$ for (x,y) such that $d(x,y) \le 0.14$. Reasoning as in part (a), we take $M_Q = 1.151$ and so we get

$$|E_Q(x,y)| \le 0.0047.$$

(c)

$$E_L(0.1, 0.1) = f(0.1, 0.1) - L(0.1, 0.1) = 1.0002 - 1 = 0.0002$$
$$E_Q(0.1, 0.1) = f(0.1, 0.1) - L(0.1, 0.1) = 1.0002 - 1 = 0.0002$$

Note: Although we want to bound E_L (respectively E_Q) on the square $|x| \le 0.1$ and $|y| \le 0.1$, we need to compute M_L (respectively M_Q) on the disk containing the above square (i.e., the disk $d(x,y) \le d_0 = 0.14$).

Solutions for Section 13.10

1. (a) The contour diagram for $f(x,y) = \frac{x}{y} + \frac{y}{x}$ is shown in Figure 13.10

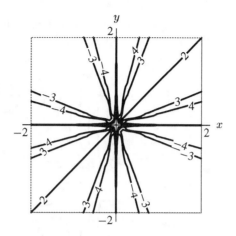

Figure 13.10

(b) If $x \neq 0$ and $y \neq 0$ then f is differentiable at (x,y). Now we need to look at points of the form $(x,0)$, where $x \neq 0$ and $(0,y)$, where $y \neq 0$. The function f is not differentiable at these points as it is not continuous.

(c) For $x \neq 0$ and $y \neq 0$,

$$f_x(x,y) = \frac{1}{y} - \frac{y}{x^2}.$$

So f_x exists for $x \neq 0$, $y \neq 0$, and it is continuous.

For all points $(x_0, 0)$ on the x-axis we have:

$$f_x(x_0, 0) = \lim_{x \to x_0} \frac{f(x,0) - f(x_0,0)}{x - x_0} = 0.$$

Thus, f_x exists but is not continuous at these points.

For points $(0, y_0)$ on the y-axis we have:

$$\lim_{x \to 0} \frac{f(x, y_0) - f(0, y_0)}{x} = \lim_{x \to 0} \left(\frac{1}{y_0} + \frac{y_0}{x^2} \right).$$

This limit doesn't exist, so the partial derivative $f_x(0, y_0)$ doesn't exist.

Similarly, for $x \neq 0$ and $y \neq 0$,

$$f_y(x,y) = -\frac{x}{y^2} + \frac{1}{x}.$$

For points $(0, y_0)$ on the y-axis we have $f_y(0, y_0) = 0$, while $f_y(x_0, 0)$ doesn't exist for $x_0 \neq 0$.

Both $f_x(x,y)$ and $f_y(x,y)$ are continuous at (x,y) only for $x \neq 0$ and $y \neq 0$.

(d) We claim f is not continuous at $(0,0)$. Let $x = t$ and $y = t$, where $t \to 0, t \neq 0$. Then

$$f(x,y) = f(t,t) = 2, \qquad \text{for} \quad t \neq 0.$$

So,

$$\lim_{t \to 0} f(t,t) = 2 \neq f(0,0) = 0,$$

and therefore

$$\lim_{(x,y) \to (0,0)} f(x,y) \neq f(0,0).$$

Thus, f is not differentiable at $(0,0)$ since f is not continuous at $(0,0)$.

(e) From part (c) we have $f_x(0,0) = 0$ and $f_y(0,0) = 0$. The functions f_x and f_y are not continuous at $(0,0)$.

5. (a)

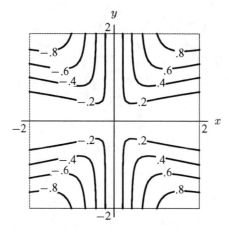

Figure 13.11

(b) f is differentiable at all $(x,y) \neq (0,0)$ as it is a rational function with nonvanishing denominator.

(c)

$$f_x(0,0) = \lim_{x \to 0} \frac{f(x,0) - f(0,0)}{x} = 0$$

$$f_y(0,0) = \lim_{y \to 0} \frac{f(0,y) - f(0,0)}{y} = 0$$

(d) Let us use the definition. If f were differentiable, the linear approximation of f would be $L(x,y) = mx + ny$, where $m = f_x(0,0) = 0$ and $n = f_y(0,0) = 0$. So let's compute

$$\lim_{(x,y) \to (0,0)} \frac{f(x,y) - L(x,y)}{\sqrt{x^2 + y^2}} = \lim_{(x,y) \to (0,0)} \frac{xy^2}{(x^2 + y^2)^{3/2}}.$$

This limit is not zero as, for $x = y = t \to 0, t > 0,$

$$\frac{xy^2}{(x^2 + y^2)^{3/2}} = \frac{t^3}{2\sqrt{2}|t|^3} \xrightarrow[t > 0]{t \to 0} \frac{1}{2\sqrt{2}}.$$

Hence f is not differentiable at $(0,0)$.

(e)

$$g(t) = f(x(t), y(t)) = \frac{ab^2 t^3}{(a^2 + b^2)t^2} = \frac{ab^2}{a^2 + b^2} t$$

So

$$g'(0) = \frac{ab^2}{a^2 + b^2}.$$

(f) $f_x(0,0) \cdot x'(0) + f_y(0,0) \cdot y'(0) = 0$, as $f_x(0,0) = f_y(0,0) = 0$. Suppose the chain rule holds, then

$$g'(t) = f_x(x(t), y(t)) \cdot x'(t) + f_x(x(t), y(t))y'(t).$$

But $g'(0) = \frac{ab^2}{a^2+b^2}$ from part (e) and $g'(0) \neq 0$ since $a \neq 0$ and $b \neq 0$. Hence the chain rule doesn't hold. This happens because f was not differentiable at $(0,0)$.

(g) If $\vec{u} = a\vec{i} + b\vec{j}$, then $a^2 + b^2 = 1$ as $\vec{u}$ is a unit vector. Thus,

$$f_{\vec{u}}(0,0) = \lim_{t \to 0} \frac{f(at, bt)}{t} = \lim_{t \to 0} \frac{g(t)}{t} = g'(0) = \frac{a^2 b}{a^2 + b^2} = a^2 b.$$

9. (a) If f were differentiable at $(0,0)$, then

$$f_{\vec{u}}(0,0) = \text{grad } f(0,0) \cdot \vec{u} = f_x(0,0) \cdot \frac{1}{\sqrt{2}} + f_y(0,0) \cdot \frac{1}{\sqrt{2}} = 0$$

which contradict the information that $f_{\vec{u}}(0,0) = 3$.

(b) Let

$$f(x, y) = \begin{cases} \frac{3}{\sqrt{2}} \left(\frac{x^2}{y} + \frac{y^2}{x} \right), & x \neq 0 \text{ and } y \neq 0, \\ 0, & x = 0 \text{ or } y = 0. \end{cases}$$

Then $f_x(0,0) = 0$, $f_y(0,0) = 0$:

$$f_{\vec{u}}(0,0) = \lim_{t \to 0} \frac{f(\frac{t}{\sqrt{2}}, \frac{t}{\sqrt{2}}) - 0}{t} = \lim_{t \to 0} \frac{3}{\sqrt{2}t} \left(\frac{t^2}{2} \cdot \frac{\sqrt{2}}{t} + \frac{t^2}{2} \cdot \frac{\sqrt{2}}{t} \right) = 3.$$

Solutions for Chapter 13 Review

1. $\dfrac{\partial z}{\partial x} = \dfrac{\partial}{\partial x} \left[(x^2 + x - y)^7 \right] = 7(x^2 + x - y)^6 (2x + 1) = (14x + 7)(x^2 + x - y)^6.$

   $\dfrac{\partial z}{\partial y} = \dfrac{\partial}{\partial y} \left[(x^2 + x - y)^7 \right] = -7(x^2 + x - y)^6.$

5.

$$\frac{\partial z}{\partial x} = 4x^3 - 7x^6 y^3 + 5y^2, \quad \frac{\partial z}{\partial y} = -3x^7 y^2 + 10xy.$$

9. Since $\|\vec{v}\| = 5$, we see that $\vec{v}$ is not a unit vector. The unit vector $\vec{u}$ in the direction of $\vec{v}$ is

$$\vec{u} = \frac{\vec{v}}{\|\vec{v}\|} = \frac{4}{5}\vec{i} - \frac{3}{5}\vec{j}.$$

The partial derivatives are $f_x(x, y) = 2xy$ and $f_y(x, y) = x^2$. So

$$f_{\vec{u}}(2,6) = f_x(2,6) \cdot \left(\frac{4}{5} \right) + f_y(2,6) \cdot \left(-\frac{3}{5} \right) = 24 \left(\frac{4}{5} \right) + 4 \left(-\frac{3}{5} \right) = \frac{84}{5}.$$

13. True. Take the direction perpendicular to grad f at that point. If grad $f = 0$, any direction will do.

17. (a) The difference quotient for evaluating $f_w(2, 2)$ is

$$f_w(2, 2) \approx \frac{f(2 + 0.01, 2) - f(2, 2)}{h} = \frac{e^{(2.01)\ln 2} - e^{2\ln 2}}{0.01} = \frac{e^{\ln(2^{2.01})} - e^{\ln(2^2)}}{0.01}$$

$$= \frac{2^{(2.01)} - 2^2}{0.01} \approx 2.78$$

The difference quotient for evaluating $f_z(2, 2)$ is

$$f_z(2, 2) \approx \frac{f(2, 2 + 0.01) - f(2, 2)}{h}$$

$$= \frac{e^{2\ln(2.01)} - e^{2\ln 2}}{0.01} = \frac{(2.01)^2 - 2^2}{0.01} = 4.01$$

(b) Using the derivative formulas we get

$$f_w = \frac{\partial f}{\partial w} = \ln z \cdot e^{w \ln z} = z^w \cdot \ln z$$

$$f_z = \frac{\partial f}{\partial z} = e^{w \ln z} \cdot \frac{w}{z} = w \cdot z^{w-1}$$

so

$$f_w(2, 2) = 2^2 \cdot \ln 2 \approx 2.773$$
$$f_z(2, 2) = 2 \cdot 2^{2-1} = 4.$$

21. The derivative $\partial c / \partial x = b$ is the rate of change of the cost of producing one unit of the product with respect to the amount of labor used (in man hours) when the amount of raw material used stays the same. Thus $\partial c / \partial x = b$ represents the hourly wage.

25. Average productivity increases as x_1 increases if $\frac{\partial}{\partial x_1}$ (average productivity) > 0. Now

$$\frac{\partial}{\partial x_1}(\text{average productivity}) = \frac{\partial}{\partial x_1}\left(\frac{P}{x_1}\right)$$

$$= \frac{1}{x_1}\frac{\partial P}{\partial x_1} + P\frac{\partial}{\partial x_1}\left(\frac{1}{x_1}\right)$$

$$= \frac{1}{x_1}\frac{\partial P}{\partial x_1} - \frac{P}{x_1^2}$$

$$= \frac{1}{x_1}\left(\frac{\partial P}{\partial x_1} - \frac{P}{x_1}\right)$$

So $\frac{\partial}{\partial x_1}$ (average productivity) > 0 means that $\left(\frac{\partial P}{\partial x_1} - \frac{P}{x_1}\right) > 0$, i.e.,

$$\frac{\partial P}{\partial x_1} > \frac{P}{x_1}.$$

29. $f_{\vec{u}} = \nabla f \cdot (\frac{1}{\sqrt{5}}\vec{i} + \frac{2}{\sqrt{5}}\vec{j})$, where $\vec{u}$ = unit vector $\frac{1}{\sqrt{5}}\vec{i} + \frac{2}{\sqrt{5}}\vec{j}$.

Now $\nabla f = \nabla(xe^y) = e^y\vec{i} + xe^y\vec{j}$.

Thus $\nabla f \cdot (\frac{1}{\sqrt{5}}\vec{i} + \frac{2}{\sqrt{5}}\vec{j}) = (e^y\vec{i} + xe^y\vec{j}) \cdot (\frac{1}{\sqrt{5}}\vec{i} + \frac{2}{\sqrt{5}}\vec{j}) = \frac{(e^y + 2xe^y)}{\sqrt{5}} = \frac{e^y}{\sqrt{5}}(2x + 1)$.

At the point $(1, 1)$, we have $f_{\vec{u}} = \frac{e^1}{\sqrt{5}}[2(1) + 1] = \frac{3e}{\sqrt{5}}$.

33. In other words, find points on the level surface $F(x, y, z) = x^2 + y^2 + z^2 = 8$ (which happens to describe a sphere of radius $2\sqrt{2}$ centered on the origin) where the normal to the sphere is parallel to $\vec{i} - \vec{j} + 3\vec{k}$, which is the normal vector of the plane $x - y + 3z = 0$. The normal to the surface is given by $\nabla F = 2x\vec{i} + 2y\vec{j} + 2z\vec{k}$. To find points (x, y, z) where the vector $2x\vec{i} + 2y\vec{j} + 2z\vec{k}$ is parallel to (i.e. a scalar multiple of) $\vec{i} - \vec{j} + 3\vec{k}$, solve $2x\vec{i} + 2y\vec{j} + 2z\vec{k} = \lambda(\vec{i} - \vec{j} + 3\vec{k})$ for λ. This equation implies that $x = \lambda/2$, $y = -\lambda/2$, and $z = 3\lambda/2$. We have

$$x^2 + y^2 + z^2 = 8$$
$$\left(\frac{\lambda}{2}\right)^2 + \left(\frac{-\lambda}{2}\right)^2 + \left(\frac{3\lambda}{2}\right)^2 = 8$$
$$\frac{\lambda^2 + \lambda^2 + 9\lambda^2}{4} = 8$$
$$11\lambda^2 = 32$$
$$\lambda = \pm\sqrt{\frac{32}{11}} = \pm\frac{4\sqrt{2}}{\sqrt{11}}$$

This gives us

$$(x, y, z) = \pm 4\sqrt{\frac{2}{11}}\left(\frac{1}{2}, -\frac{1}{2}, \frac{3}{2}\right)$$

37. (a) $\dfrac{\partial w}{\partial u} = \dfrac{\partial w}{\partial x} \cdot \dfrac{\partial x}{\partial u} + \dfrac{\partial w}{\partial y} \cdot \dfrac{\partial y}{\partial u} + \dfrac{\partial w}{\partial z} \cdot \dfrac{\partial z}{\partial u} = 3y \cdot \dfrac{1}{u} + (3x + z) \cdot \sin v + y \cdot v.$

At $(u, v) = (1, \pi)$ we have $y = 1, x = -1, z = \pi$.

Thus, $\dfrac{\partial w}{\partial u}\Big|_{(1,\pi)} = 3 + (-3 + \pi) \cdot 0 + \pi = 3 + \pi.$

$\dfrac{\partial w}{\partial v} = \dfrac{\partial w}{\partial x} \cdot \dfrac{\partial x}{\partial v} + \dfrac{\partial w}{\partial y} \cdot \dfrac{\partial y}{\partial v} + \dfrac{\partial w}{\partial z} \cdot \dfrac{\partial z}{\partial v} = 3y(-\sin v) + (3x + z) \cdot u\cos v + yu.$

Thus, $\dfrac{\partial w}{\partial v}\Big|_{(1,\pi)} = 3(1)(0) + (-3 + \pi)1(-1) + 1(1) = 4 - \pi.$

(b) $\dfrac{dw}{dt} = \dfrac{\partial w}{\partial u} \cdot \dfrac{du}{dt} + \dfrac{\partial w}{\partial v} \cdot \dfrac{dv}{dt}.$

$\dfrac{du}{dt} = \pi\cos\pi t$, so $\dfrac{du}{dt}\Big|_{t=1} = -\pi$ and $\dfrac{dv}{dt} = 2\pi t$, so $\dfrac{dv}{dt}\Big|_{t=1} = 2\pi.$

Thus $\dfrac{dw}{dt}\Big|_{t=1} = (3 + \pi)(-\pi) + (4 - \pi)2\pi = 5\pi - 3\pi^2.$

41. Let us first collect the computations that we will need.

$$f(x, y) = \cos(x + 2y)\sin(x - y),$$
$$f_x(x, y) = \cos(x + 2y)\cos(x - y) - \sin(x - y)\sin(x + 2y)$$
$$= \cos(x + 2y + x - y) = \cos(2x + y),$$
$$f_y(x, y) = -\cos(x + 2y)\cos(x - y) - 2\sin(x - y)\sin(x + 2y)$$

$$= -\cos\left(x + 2y - (x - y)\right) - \sin\left(x + 2y\right)\sin\left(x - y\right)$$
$$= -\cos\left(3y\right) + \frac{1}{2}\left[\cos\left(2x + y\right) - \cos\left(3y\right)\right]$$
$$= \frac{1}{2}\cos\left(2x + y\right) - \frac{3}{2}\cos\left(3y\right),$$
$$f_{xx}(x, y) = -2\sin\left(2x + y\right),$$
$$f_{xy}(x, y) = -\sin\left(2x + y\right),$$
$$f_{yy}(x, y) = -\frac{1}{2}\sin\left(2x + y\right) + \frac{9}{2}\sin\left(3y\right).$$

Then

$$f(0,0) = 0,$$
$$f_x(0,0) = 1,$$
$$f_y(0,0) = -1,$$
$$f_{xx}(0,0) = 0,$$
$$f_{xy}(0,0) = 0,$$
$$f_{yy}(0,0) = 0.$$

Hence the quadratic Taylor polynomial $P(x, y)$ of $f(x, y)$ at $(0, 0)$ is

$$P(x, y) = f(0,0) + f_x(0,0)x + f_y(0,0)y$$
$$+ \frac{1}{2}f_{xx}(0,0)x^2 + f_{xy}(0,0)xy + \frac{1}{2}f_{yy}(0,0)y^2$$
$$= x - y.$$

CHAPTER FOURTEEN

Solutions for Section 14.1

1. The point A is not a critical point and the contour lines look like parallel lines. The point B is a critical point and is a local maximum; the point C is a saddle point.

5. The partial derivatives are $f_x(x, y) = 3x^2 - 3$ which vanishes for $x = \pm 1$ and $f_y(x, y) = 3y^2 - 3$ which vanishes for $y = \pm 1$. The points $(1, 1), (1, -1), (-1, 1), (-1, -1)$ where both partials vanish are the critical points. To determine the nature of these critical points we calculate their discriminant and use the second derivative test. The discriminant is

$$D = f_{xx}(x, y)f_{yy}(x, y) - f_{xy}^2(x, y) = (6x)(6y) - 0 = 36xy.$$

At $(1, 1)$ and $(-1, -1)$ the discriminant is positive. Since $f_{xx}(1, 1) = 6$ is positive, $(1, 1)$ is a local minimum. And since $f_{xx}(-1, -1) = -6$ is negative, $(-1, -1)$ is a local maximum. The remaining two points, $(1, -1)$ and $(-1, 1)$ are saddle points since the discriminant is negative.

9. To find critical points, set partial derivatives equal to zero:

$$E_x = \sin x = 0 \quad \text{when} \quad x = 0, \pm\pi, \pm 2\pi, \cdots$$

$$E_y = y = 0 \quad \text{when} \quad y = 0.$$

The critical points are

$$\cdots (-2\pi, 0), (-\pi, 0), (0, 0), (\pi, 0), (2\pi, 0), (3\pi, 0) \cdots$$

To classify, calculate $D = E_{xx}E_{yy} - (E_{xy})^2 = \cos x$.
At the points $(0, 0), (\pm 2\pi, 0), (\pm 4\pi, 0), (\pm 6\pi, 0), \cdots$

$$D = (1) > 0 \quad \text{and} \quad E_{xx} > 0 \quad (\text{Since} E_{xx}(0, 2k\pi) = \cos(2k\pi) = 1).$$

Therefore $(0, 0), (\pm 2\pi, 0), (\pm 4\pi, 0), (\pm 6\pi, 0), \cdots$ are local minima.
At the points $(\pm\pi, 0), (\pm 3\pi, 0), (\pm 5\pi, 0), (\pm 7\pi, 0), \cdots$, we have $\cos(2k + 1)\pi = -1$, so

$$D = (-1) < 0.$$

Therefore $(\pm\pi, 0), (\pm 3\pi, 0), (\pm 5\pi, 0), (\pm 7\pi, 0), \cdots$ are saddle points.

13. At the origin $f(0, 0) = 0$. Since $x^6 \geq 0$ and $y^6 \geq 0$, the point $(0, 0)$ is a local (and global) minimum. The second derivative test does not tell you anything since $D = 0$.

17. (a) $(1, 3)$ is a critical point. Since $f_{xx} > 0$ and the discriminant

$$D = f_{xx}f_{yy} - f_{xy}^2 = f_{xx}f_{yy} - 0^2 = f_{xx}f_{yy} > 0,$$

the point $(1, 3)$ is a minimum.

(b)

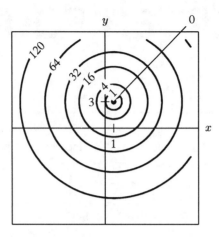

Figure 14.1

Solutions for Section 14.2

1. Mississippi lies entirely within a region designated as "80s" so we expect both the high and low daily temperatures within the state to be in the 80s. The South-Western most corner of the state is close to a region designated as "90s" so, we would expect the temperature here to be in the high 80s, say 87-88. The northern most portion of the state is relatively central to the "80s" region. We might expect the temperature there to be between 83-87.

Alabama also lies completely within a region designated as "80s" so both the high and low daily temperatures within the state are in the 80s. The south-eastern tip of the state is close to a "90s" region so we would expect the temperature here to be ≈ 88-89 degrees. The northern most part of the state is relatively central to the "80s" region so the temperature there is ≈ 83-87 degrees.

Pennsylvania is also in the "80s" region, but it is touched by the boundary line between the "80s" and a "70s" region. Thus we expect the low daily temperature to occur there and be about 70 degrees. The state is also touched by a boundary line which contains a "90s" region so the high will occur there and be 89-90 degrees.

New York is split by a boundary between an "80s" and a "70s" region, the northern portion of the state is apt to be about 74-76 while the southern portion is likely to be in the low 80s, maybe 81-84 or so.

California contains many different zones. The northern coastal areas will probably have the state daily low at 65-68, although without another contour on that side, it is difficult to judge how quickly the temperature is dropping off to the west. The tip of Southern California is in a 100s region, so there we expect the state daily high to be 100-101.

Arizona will have a low around 85-87 in the northwest corner and a high in the 100s, perhaps 102-107 in its southern regions.

Massachusetts will probably have a high around 81-84 and a low at 70.

5. To maximize $z = x^2 + y^2$, it suffices to maximize x^2 and y^2. We can maximize both of these at the same time by taking the point $(1, 1)$, where $z = 2$. It occurs on the boundary of the square. (Note: We also have maxima at the points $(-1, -1), (-1, 1)$ and $(1, -1)$ which are on the boundary of the square.)

To minimize $z = x^2 + y^2$, we choose the point $(0, 0)$, where $z = 0$. It does not occur on the boundary of the square.

9. The total revenue is
$$R = pq = (60 - 0.04q)q = 60q - 0.04q^2,$$

and as $q = q_1 + q_2$, this gives
$$R = 60q_1 + 60q_2 - 0.04q_1^2 - 0.08q_1q_2 - 0.04q_2^2.$$

Therefore, the profit is
$$\begin{aligned} P(q_1, q_2) &= R - C_1 - C_2 \\ &= -13.7 + 60q_1 + 60q_2 - 0.07q_1^2 - 0.08q_2^2 - 0.08q_1q_2. \end{aligned}$$

At a local maximum point, we have grad $P = \vec{0}$:
$$\frac{\partial P}{\partial q_1} = 60 - 0.14q_1 - 0.08q_2 = 0,$$
$$\frac{\partial P}{\partial q_2} = 60 - 0.16q_2 - 0.08q_1 = 0.$$

Solving these equations, we find that
$$q_1 = 300 \quad \text{and} \quad q_2 = 225.$$

To see whether or not we have found a local maximum, we compute the second-order partial derivatives:
$$\frac{\partial^2 P}{\partial q_1^2} = -0.14, \quad \frac{\partial^2 P}{\partial q_2^2} = -0.16, \quad \frac{\partial^2 P}{\partial q_1 \partial q_2} = -0.08.$$

Therefore,
$$D = \frac{\partial^2 P}{\partial q_1^2}\frac{\partial^2 P}{\partial q_2^2} - \frac{\partial^2 P}{\partial q_1 \partial q_2} = (-0.14)(-0.16) - (-0.08)^2 = 0.016,$$

and so we have found a local maximum point. The graph of $P(q_1, q_2)$ has the shape of an upside down paraboloid since P is quadratic in q_1 and q_2, hence $(300, 225)$ is a global maximum point.

13. The function $f(x, y)$ in Example 3 is given by
$$f(x, y) = \frac{80}{xy} + 10x + 10xy + 20y.$$

This has critical points when $f_x = f_y = 0$.
$$f_x(x, y) = \frac{-80}{x^2 y} + 10 + 10y.$$

$$f_y(x, y) = \frac{-80}{xy^2} + 10x + 20.$$

Substituting $x = 2, y = 1$ gives
$$f_x(2, 1) = \frac{-80}{2^2 \cdot 1} + 10 + 10.1 = 0$$
$$f_y(2, 1) = \frac{-80}{2 \cdot 1^2} + 10.2 + 20 = 0.$$

So our point, $(2, 1)$, is a critical point.

To determine if this critical point is a minimum we use the second derivative test.

$$f_{xx} = \frac{160}{x^3 y}, \qquad f_{xx}(2, 1) = 20,$$
$$f_{yy} = \frac{160}{x y^3}, \qquad f_{yy}(2, 1) = 80,$$
$$f_{xy} = \frac{80}{x^2 y^2} + 10, \qquad f_{xy}(2, 1) = 30.$$

So $D = 20 \cdot 80 - 30^2 = 700 > 0$ and $f_{xx}(2, 1) > 0$, therefore the point $(2, 1)$ is a local minimum.

17. Let the line be in the form $y = b + mx$. When x equals -1, 0 and 1, then y equals $b - m$, b, and $b + m$, respectively. The sum of the squares of the vertical distances, which is what we want to minimize, is

$$f(m, b) = (2 - (b - m))^2 + (-1 - b)^2 + (1 - (b + m))^2.$$

To find the critical points, we compute the partial derivatives with respect to m and b,

$$\begin{aligned} f_m &= 2(2 - b + m) + 0 + 2(1 - b - m)(-1) \\ &= 4 - 2b + 2m - 2 + 2b + 2m \\ &= 2 + 4m, \\ f_b &= 2(2 - b + m)(-1) + 2(-1 - b)(-1) + 2(1 - b - m)(-1) \\ &= -4 + 2b - 2m + 2 + 2b - 2 + 2b + 2m \\ &= -4 + 6b. \end{aligned}$$

Setting both partial derivatives equal to zero, we get a system of equations:

$$2 + 4m = 0,$$
$$-4 + 6b = 0.$$

The solution is $m = -1/2$ and $b = 2/3$. One can check that it is a minimum. Hence, the regression line is $y = \frac{2}{3} - \frac{1}{2}x$.

21. (a) We have the following contour diagram for f:

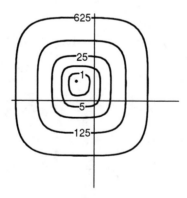

Figure 14.2

(b) We first compute grad f:

$$f_x = 4(x + 1)^3 - \frac{2xy^2}{(x^2 y^2 + 1)^2} \quad \text{and} \quad f_y = 4(y - 1)^3 - \frac{2yx^2}{(x^2 y^2 + 1)^2}.$$

These equations are difficult to solve simultaneously and this is why we need to use a gradient search method. We choose $(x_0, y_0) = (-1, 1)$ as our starting point and compute

$$\text{grad } f(-1, 1) = 0.5\vec{i} - 0.5\vec{j}.$$

Since we wish to minimize f, we move from (x_0, y_0) in the opposite direction of the gradient to a point

$$(x_1(t), y_1(t)) = (x_0, y_0) - t \text{ grad } f(x_0, y_0) = (-1, 1) - t(0.5, -0.5),$$

such that $f(x_1, y_1) < f(x_0, y_0)$. We should choose t so that the function $f(x_1(t), y_1(t))$ is minimized. Since the function $f(x, y)$ is complicated and only an approximate answer is required, we will try $t = 0.5$ for each iteration of the gradient search method. Therefore,

$$(x_1, y_1) = (-1.25, 1.25),$$

$$\text{grad } f(-1.25, 1.25) \approx 0.27\vec{i} - 0.27\vec{j}.$$

Repeating this step, again with $t = 0.5$, gives

$$(x_2, y_2) = (-1.25, 1.25) - (0.5)(0.27, -0.27) \approx (-1.38, 1.38),$$

$$\text{grad } f(-1.38, 1.38) \approx 0.02\vec{i} - 0.02\vec{j}.$$

Performing one more iteration, we get

$$(x_3, y_3) = (-1.38, 1.38) - (0.5)(0.02, -0.02) \approx (-1.39, 1.39),$$

$$\text{grad } f(-1.39, 1.39) \approx 0.002\vec{i} - 0.002\vec{j},$$

so we are already very close to a critical point. We find that $f(-1.39, 1.39) \approx 0.2575$ and verify that this is a global minimum using the contour diagram from part (a).

Solutions for Section 14.3

1. Our objective function is $f(x, y) = x + y$ and our equation of constraint is $g(x, y) = x^2 + y^2 = 1$. To optimize $f(x, y)$ with Lagrange multipliers, we solve $\nabla f(x, y) = \lambda \nabla g(x, y)$ subject to $g(x, y) = 1$. The gradients of f and g are

$$\nabla f(x, y) = \vec{i} + \vec{j},$$
$$\nabla g(x, y) = 2x\vec{i} + 2y\vec{j}.$$

So the equation $\nabla f = \lambda \nabla g$ becomes

$$\vec{i} + \vec{j} = \lambda(2x\vec{i} + 2y\vec{j})$$

Solving for λ gives

$$\lambda = \frac{1}{2x} = \frac{1}{2y},$$

which tells us that $x = y$. Going back to our equation of constraint, we use the substitution $x = y$ to solve for y

$$g(y, y) = y^2 + y^2 = 1$$
$$2y^2 = 1$$
$$y^2 = \frac{1}{2}$$
$$y = \pm\sqrt{\frac{1}{2}} = \pm\frac{\sqrt{2}}{2}.$$

Since $x = y$, our critical points are $(\frac{\sqrt{2}}{2}, \frac{\sqrt{2}}{2})$ and $(-\frac{\sqrt{2}}{2}, -\frac{\sqrt{2}}{2})$. Evaluating f at these points we find that the maximum value is $f(\frac{\sqrt{2}}{2}, \frac{\sqrt{2}}{2}) = \sqrt{2}$ and the minimum value is $f(-\frac{\sqrt{2}}{2}, -\frac{\sqrt{2}}{2}) = -\sqrt{2}$.

5. The objective function is $f(x, y) = x^2 + y^2$ and the equation of constraint is $g(x, y) = x^4 + y^4 = 2$. Their gradients are

$$\nabla f(x, y) = 2x\vec{i} + 2y\vec{j},$$
$$\nabla g(x, y) = 4x^3\vec{i} + 4y^3\vec{j}.$$

So the equation $\nabla f = \lambda \nabla g$ becomes $2x\vec{i} + 2y\vec{j} = \lambda(4x^3\vec{i} + 4y^3\vec{j})$. This tells us that

$$2x = 4\lambda x^3,$$
$$2y = 4\lambda y^3.$$

Now if $x = 0$, the first equation is true for any value of λ. In particular, we can choose λ which satisfies the second equation. Similarly, $y = 0$ is solution.

Assuming both $x \neq 0$ and $y \neq 0$, we can divide to solve for λ and find

$$\lambda = \frac{2x}{4x^3} = \frac{2x}{4y^3}$$
$$\frac{1}{2x^2} = \frac{1}{2y^2}$$
$$y^2 = x^2$$
$$y = \pm x.$$

Going back to our equation of constraint, we find

$$g(0, y) = 0^4 + y^4 = 2 \Rightarrow y = \pm\sqrt[4]{2}$$
$$g(x, 0) = x^4 + 0^4 = 2 \Rightarrow x = \pm\sqrt[4]{2}$$
$$g(x, \pm x) = x^4 + (\pm x)^4 = 2 \Rightarrow x = \pm 1.$$

Thus, the critical points are $(0, \pm\sqrt[4]{2})$, $(\pm\sqrt[4]{2}, 0)$, $(1, \pm 1)$ and $(-1, \pm 1)$. Evaluating f at these points we find

$$f(1, 1) = f(1, -1) = f(-1, 1) = f(-1, -1) = 2,$$
$$f(0, \sqrt[4]{2}) = f(0, -\sqrt[4]{2}) = f(\sqrt[4]{2}, 0) = f(-\sqrt[4]{2}, 0) = \sqrt{2}.$$

So the minimum value of $f(x, y)$ on $g(x, y) = 2$ is $\sqrt{2}$ and the maximum value is 2.

9. Our objective function is $f(x, y, z) = x^2 - y^2 - 2z$ and our equation of constraint is $g(x, y, z) = x^2 + y^2 - z = 0$. To optimize $f(x, y, z)$ with Lagrange multipliers, we solve $\nabla f(x, y, z) = \lambda \nabla g(x, y, z)$ subject to $g(x, y, z) = 0$. The gradients of f and g are

$$\nabla f(x, y, z) = 2x\vec{i} - 2y\vec{j} - 2\vec{k},$$
$$\nabla g(x, y, z) = 2x\vec{i} + 2y\vec{j} - \vec{k}.$$

We get

$$2x = 2\lambda x$$
$$-2y = 2\lambda y$$
$$-2 = -\lambda$$
$$x^2 + y^2 = z.$$

The third equation gives $\lambda = 2$ and from the first $x = 0$, from the second $y = 0$ and from the fourth $z = 0$. So the only solution is $(0, 0, 0)$, and $f(0, 0, 0) = 0$.

To see what kind of extreme point is $(0, 0, 0)$, let (a, b, c) be a point which satisfies the constraint, i.e. $a^2 + b^2 = c$. Then $f(a, b, c) = a^2 - b^2 - 2c = -a^2 - 3b^2 \leq 0$. The conclusion is that 0 is the maximum value of f and that there is no minimum.

13. The domain $x^2 + 2y^2 \leq 1$ is the shaded interior of the ellipse $x^2 + 2y^2 = 1$ including the boundary, shown in Figure 14.3.

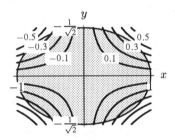

Figure 14.3

First we want to find the relative maxima and minima of f in the interior of the ellipse. So we need to find the extrema of

$$f(x, y) = xy, \quad \text{in the region} \quad x^2 + 2y^2 < 1.$$

For this we compute the critical points:

$$f_x = y = 0 \quad \text{and} \quad f_y = x = 0.$$

So there is one critical point, $(0, 0)$. As $f_{xx}(0, 0) = 0$, $f_{yy}(0, 0) = 0$ and $f_{xy}(0, 0) = 1$ we have

$$D = f_{xx}(0, 0) \cdot f_{yy}(0, 0) - (f_{xy}(0, 0))^2 = -1 < 0$$

so $(0, 0)$ is a saddle and f doesn't have relative extrema in the interior of the ellipse.

Now let's find the relative extrema of f on the boundary, hence this time we'll have a constraint problem. We want the extrema of $f(x, y) = xy$ subject to $g(x, y) = x^2 + 2y^2 - 1 = 0$. We use Lagrange multipliers:

$$\text{grad } f = \lambda \text{ grad } g \quad \text{and} \quad x^2 + 2y^2 = 1$$

which give

$$y = 2\lambda x$$
$$x = 4\lambda y$$
$$x^2 + 2y^2 = 1$$

From the first two equations we get

$$xy = 8\lambda^2 xy.$$

So $x = 0$ or $y = 0$ or $8\lambda^2 = 1$.

If $x = 0$, from the last equation $2y^2 = 1$ so $y = \pm\frac{\sqrt{2}}{2}$ and we get the solutions $(0, \frac{\sqrt{2}}{2})$ and $(0, -\frac{\sqrt{2}}{2})$.

If $y = 0$, from the last equation we get $x^2 = 1$ and so the solutions are $(1, 0)$ and $(-1, 0)$.

If $x \neq 0$ and $y \neq 0$ then $8\lambda^2 = 1$, hence $\lambda = \pm\frac{1}{2\sqrt{2}}$. For $\lambda = \frac{1}{2\sqrt{2}}$

$$x = \sqrt{2}y$$

and plugging into the third equation gives $4y^2 = 1$ so we get the solutions $(\frac{\sqrt{2}}{2}, \frac{1}{2})$ and $(-\frac{\sqrt{2}}{2}, -\frac{1}{2})$.

For $\lambda = -\frac{1}{2\sqrt{2}}$ we get

$$x = -\sqrt{2}y$$

and plugging into the third equation gives $4y^2 = 1$, and the solutions $(\frac{\sqrt{2}}{2}, -\frac{1}{2})$ and $(-\frac{\sqrt{2}}{2}, \frac{1}{2})$. So finally we have the solutions: $(1, 0)$, $(-1, 0)$, $(\frac{\sqrt{2}}{2}, \frac{1}{2})$, $(-\frac{\sqrt{2}}{2}, -\frac{1}{2})$, $(\frac{\sqrt{2}}{2}, -\frac{1}{2})$, $(-\frac{\sqrt{2}}{2}, \frac{1}{2})$.

Evaluating f at these points gives:

$$f(0, \frac{\sqrt{2}}{2}) = f(0, -\frac{\sqrt{2}}{2}) = f(1, 0) = f(-1, 0) = 0$$
$$f(\frac{\sqrt{2}}{2}, \frac{1}{2}) = f(-\frac{\sqrt{2}}{2}, -\frac{1}{2}) = \frac{\sqrt{2}}{4}$$
$$f(\frac{\sqrt{2}}{2}, -\frac{1}{2}) = f(-\frac{\sqrt{2}}{2}, \frac{1}{2}) = -\frac{\sqrt{2}}{4}.$$

Hence the maximum value of f is $\frac{\sqrt{2}}{4}$ and the minimum value of f is $-\frac{\sqrt{2}}{4}$.

17. The region $x^2 + y^2 \leq 1$ is the shaded disk of radius 1 centered at the origin (including the circle $x^2 + y^2 = 1$) shown in Figure 14.4.

Let's first compute the critical points of f in the interior of the disk. We have

$$f_x = 3x^2 = 0$$
$$f_y = -2y = 0,$$

whose solution is $x = y = 0$. So the only one critical point is $(0, 0)$. As $f_{xx}(0, 0) = 0$, $f_{yy}(0, 0) = -2$ and $f_{xy}(0, 0) = 0$,

$$D = f_{xx}(0, 0) \cdot f_{yy}(0, 0) - (f_{xy}(0, 0))^2 = 0$$

which doesn't tell us anything about the nature of the critical point $(0, 0)$.

But, if we choose x, y very small in absolute value and such that $x^3 > y^2$, then $f(x, y) > 0$. If we choose x, y very small in absolute value and such that $x^3 < y^2$, then $f(x, y) < 0$. As $f(0, 0) = 0$, we conclude that $(0, 0)$ is a saddle point.

We can get the same conclusion looking at the level curves of f around $(0, 0)$, as shown in Figure 14.5.

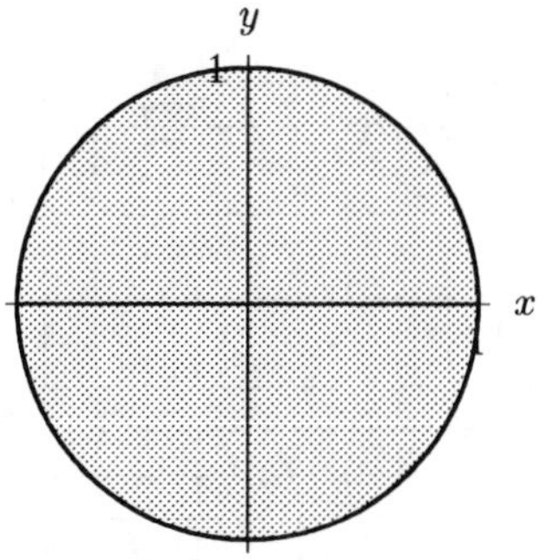

Figure 14.4

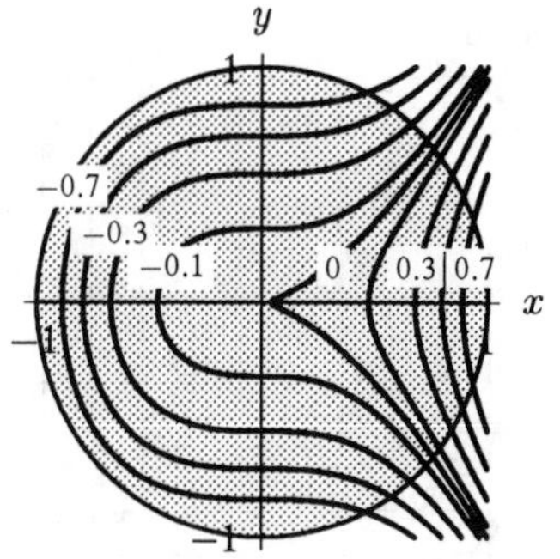

Figure 14.5: Level curves of f

So, f doesn't have extrema in the interior of the disk.

Now, let's find the relative extrema of f on the circle $x^2 + y^2 = 1$. So we want the extrema of $f(x,y) = x^3 - y^2$ subject to the constraint $g(x,y) = x^2 + y^2 - 1 = 0$. Using Lagrange multipliers we get

$$\text{grad } f = \lambda \text{ grad } g \quad \text{and} \quad x^2 + y^2 = 1,$$

which gives

$$3x^2 = 2\lambda x$$
$$-2y = 2\lambda y$$
$$x^2 + y^2 = 1.$$

From the second equation $y = 0$ or $\lambda = -1$.

If $y = 0$, from the third equation we get $x^2 = 1$, which gives the solutions $(1,0), (-1,0)$.

If $y \neq 0$ then $\lambda = -1$ and from the first equation we get $3x^2 = -2x$, hence $x = 0$ or $x = -\frac{2}{3}$. If $x = 0$, from the third equation we get $y^2 = 1$, so the solutions $(0,1),(0,-1)$. If $x = -\frac{2}{3}$, from the third equation we get $y^2 = \frac{5}{9}$, so the solutions $(-\frac{2}{3}, \frac{\sqrt{5}}{3}), (-\frac{2}{3}, -\frac{\sqrt{5}}{3})$.

Evaluating f at these points we get

$$f(1,0) = 1, \quad f(-1,0) = f(0,1) = f(0,-1) = -1$$

and

$$f\left(-\frac{2}{3}, -\frac{\sqrt{5}}{3}\right) = f\left(-\frac{2}{3}, \frac{\sqrt{5}}{3}\right) = -\frac{23}{27}.$$

Therefore the maximum value of f is 1 and the minimum value is -1.

21. (a)   The curves are shown in Figure 14.6.

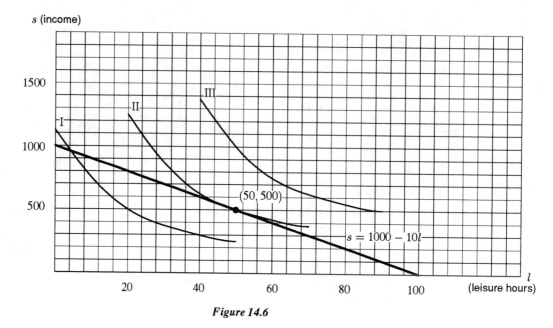

Figure 14.6

(b)   The income equals $10/hour times the number of hours of work:

$$s = 10(100 - l) = 1000 - 10l.$$

(c)   The graph of this constraint is the straight line in Figure 14.6.

(d)   For any given salary, curve III allows for the most leisure time, curve I the least. Similarly, for any amount of leisure time, any curve III also has the greatest salary, and curve I the least. Thus, any point on curve III is preferable to any point on curve II, which is preferable to any point on curve I. We prefer to be on the outermost curve that our constraint allows. We want to choose the point on $s = 1000 - 10l$ which is on the most preferable curve. Since all the curves are concave up, this occurs at the point where $s = 1000 - 10l$ is *tangent* to curve II. So we choose $l = 50$, $s = 500$, and work 50 hours a week.

25.   We want to minimize the function $h(x, y)$ subject to the constraint that

$$g(x, y) = x^2 + y^2 = 1,000^2 = 1,000,000.$$

Using the method of Lagrange multipliers, we obtain the following system of equations:

$$h_x = -\frac{10x + 4y}{10,000} = 2\lambda x,$$

$$h_y = -\frac{4x + 4y}{10,000} = 2\lambda y,$$

$$x^2 + y^2 = 1,000,000.$$

Multiplying the first equation by y and the second by x we get

$$\frac{-y(10x + 4y)}{10,000} = \frac{-x(4x + 4y)}{10,000}.$$

Hence:

$$2y^2 + 3xy - 2x^2 = (2y - x)(y + 2x) = 0,$$

and so the climber either moves along the line $x = 2y$ or $y = -2x$.

We must now choose one of these lines and the direction along that line which will lead to the point of minimum height on the circle. To do this we find the points of intersection of these lines with the circle $x^2 + y^2 = 1,000,000$, compute the corresponding heights, and then select the minimum point.

If $x = 2y$, the third equation gives

$$5y^2 = 1,000^2,$$

so that $y = \pm 1,000/\sqrt{5} \approx \pm 447.21$ and $x = \pm 894.43$. The corresponding height is $h(\pm 894.43, \pm 447.21) = 2400$ m. If $y = -2x$, we find that $x = \pm 447.21$ and $y = \mp 894.43$. The corresponding height is $h(\pm 447.21, \mp 894.43) = 2900$ m. Therefore, she should travel along the line $x = 2y$, in either of the two possible directions.

29. (a) We draw the level curves (parallel straight lines) of $f(x, y) = ax + by + c$. We can see that the level lines with the maximum and minimum f-values which intersect with the disk are the level lines that are tangent to the boundary of the disk. Therefore, the maximum and minimum occur at the boundary of the disk. See Figure 14.7.

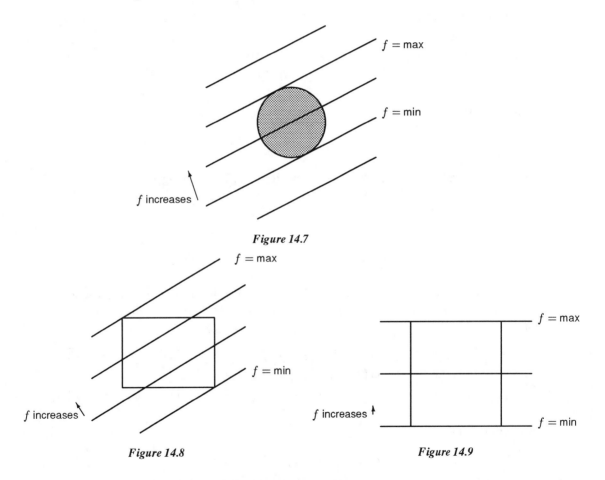

Figure 14.7

Figure 14.8

Figure 14.9

(b) Similar to part (a), we see the level lines with the largest and smallest f-values which intersect with the rectangle must pass the corner of the rectangle. So the maximum and minimum occur at the corners of rectangle. See Figure 14.8. When the level curves are parallel to a pair of the sides, then the points on the sides are all maximum or minimum, as shown below in Figure 14.9.

(c)   The graph of f is a plane. The part of the graph lying above a disk R is either a flat disk, in which case every point is a maximum, or is a tilted ellipse, in which case you can see that the maximum will be on the edge. Similarly, the part lying above a rectangle is either a rectangle or a tilted parallelogram, in which case the maximum will be at a corner.

Solutions for Chapter 14 Review

1.   The partial derivatives are

$$f_x = \cos x + \cos(x + y).$$
$$f_y = \cos y + \cos(x + y).$$

Setting $f_x = 0$ and $f_y = 0$ gives

$$\cos x = \cos y$$

For $0 < x < \pi$ and $0 < y < \pi$, $\cos x = \cos y$ only if $x = y$. Then, setting $f_x = f_y = 0$:

$$\cos x + \cos 2x = 0,$$
$$\cos x + 2\cos^2 x - 1 = 0,$$
$$(2\cos x - 1)(\cos x + 1) = 0.$$

So $\cos x = 1/2$ or $\cos x = -1$, that is $x = \pi/3$ or $x = \pi$. For the given domain $0 < x < \pi$, $0 < y < \pi$, we only consider the solution when $x = \pi/3$ then $y = x = \pi/3$. Therefore, the critical point is $(\frac{\pi}{3}, \frac{\pi}{3})$. Since

$$f_{xx}(x, y) = -\sin x - \sin(x + y) \quad f_{xx}(\frac{\pi}{3}, \frac{\pi}{3}) = -\sin\frac{\pi}{3} - \sin\frac{2\pi}{3} = -\sqrt{3}$$
$$f_{xy}(x, y) = -\sin(x + y) \qquad\qquad f_{xy}(\frac{\pi}{3}, \frac{\pi}{3}) = -\sin\frac{2\pi}{3} \qquad\quad = -\frac{\sqrt{3}}{2}$$
$$f_{yy}(x, y) = -\sin y - \sin(x + y) \quad f_{yy}(\frac{\pi}{3}, \frac{\pi}{3}) = -\sin\frac{\pi}{3} - \sin\frac{2\pi}{3} = -\sqrt{3}$$

the discriminant is

$$D(x, y) = f_{xx}f_{yy} - f_{xy}^2$$
$$= (\sqrt{3})(-\sqrt{3}) - (-\frac{\sqrt{3}}{2})^2 = \frac{9}{4} > 0.$$

Since $f_{xx}(\frac{\pi}{3}, \frac{\pi}{3}) = -\sqrt{3} < 0$, $(\frac{\pi}{3}, \frac{\pi}{3})$ is a local maximum.

5.   Since $f_{xx} < 0$ and $D = f_{xx}f_{yy} - f_{xy}^2 > 0$, $(1, 3)$ is a maximum.

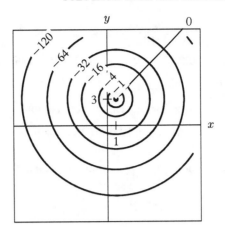

Figure 14.10

9. We first express the revenue R in terms of the prices p_1 and p_2:

$$R(p_1, p_2) = p_1 q_1 + p_2 q_2$$
$$= p_1(517 - 3.5p_1 + 0.8p_2) + p_2(770 - 4.4p_2 + 1.4p_1)$$
$$= 517p_1 - 3.5p_1^2 + 770p_2 - 4.4p_2^2 + 2.2p_1 p_2.$$

At a local maximum we have grad $R = 0$, and so:

$$\frac{\partial R}{\partial p_1} = 517 - 7p_1 + 2.2p_2 = 0,$$

$$\frac{\partial R}{\partial p_2} = 770 - 8.8p_2 + 2.2p_1 = 0.$$

Solving these equations, we find that

$$p_1 = 110 \quad \text{and} \quad p_2 = 115.$$

To see whether or not we have a found a local maximum, we compute the second-order partial derivatives:

$$\frac{\partial^2 R}{\partial p_1^2} = -7, \quad \frac{\partial^2 R}{\partial p_2^2} = -8.8, \quad \frac{\partial^2 R}{\partial p_1 \partial p_2} = 2.2.$$

Therefore,

$$D = \frac{\partial^2 R}{\partial p_1^2} \frac{\partial^2 R}{\partial p_2^2} - \frac{\partial^2 R}{\partial p_1 \partial p_2} = (-7)(-8.8) - (2.2)^2 = 56.76,$$

and so we have found a local maximum point. The graph of $P(p_1, p_2)$ has the shape of an upside down paraboloid. Since P is quadratic in q_1 and q_2, $(110, 115)$ is a global maximum point.

13. (a) To be producing the maximum quantity Q under the cost constraint given, the firm should be using K and L values given by $\nabla Q = \lambda \nabla C, C = 20K + 10L = 150$, so

$$\frac{\partial Q}{\partial K} = 0.6aK^{-0.4}L^{0.4} = 20\lambda$$

$$\frac{\partial Q}{\partial L} = 0.4aK^{0.6}L^{-0.6} = 10\lambda$$

Hence $\dfrac{0.6aK^{-0.4}L^{0.4}}{0.4aK^{0.6}L^{-0.6}} = 1.5\dfrac{L}{K} = \dfrac{20\lambda}{10\lambda} = 2$, so $L = \dfrac{4}{3}K$. Substituting in $20K + 10L = 150$, we obtain $20K + 10\left(\dfrac{4}{3}\right)K = 150$. Then $K = \dfrac{9}{2}$ and $L = 6$, so capital should be reduced by $\dfrac{1}{2}$ unit, and labor should be increased by 1 unit.

(b) $\dfrac{\text{New production}}{\text{Old production}} = \dfrac{a4.5^{0.6}6^{0.4}}{a5^{0.6}5^{0.4}} \approx 1.01$, so tell the board of directors, "Reducing the quantity of capital by $1/2$ unit and increasing the quantity of labor by 1 unit will increase production by 1% while holding costs to \$150."

17. The wetted perimeter of the trapezoid is given by the sum of the lengths of the three walls, so

$$p = w + \frac{2d}{\sin\theta}$$

We want to minimize p subject to the constraint that the area is fixed at 50 m^2. A trapezoid of height h and with parallel sides of lengths b_1 and b_2 has

$$A = \text{Area} = h\frac{(b_1 + b_2)}{2}.$$

In this case, d corresponds to h and b_1 corresponds to w. The b_2 term corresponds to the width of the exposed surface of the canal. We find that $b_2 = w + (2d)/(\tan\theta)$. Substituting into our original equation for the area along with the fact that the area is fixed at 50 m^2, we arrive at the formula:

$$\text{Area} = \frac{d}{2}\left(w + w + \frac{2d}{\tan\theta}\right) = d\left(w + \frac{d}{\tan\theta}\right) = 50$$

We now solve the constraint equation for one of the variables; we will choose w to give

$$w = \frac{50}{d} - \frac{d}{\tan\theta}.$$

Substituting into the expression for p gives

$$p = w + \frac{2d}{\sin\theta} = \frac{50}{d} - \frac{d}{\tan\theta} + \frac{2d}{\sin\theta}.$$

We now take partial derivatives:

$$\frac{\partial p}{\partial d} = -\frac{50}{d^2} - \frac{1}{\tan\theta} + \frac{2}{\sin\theta}$$

$$\frac{\partial p}{\partial\theta} = \frac{d}{\tan^2\theta}\cdot\frac{1}{\cos^2\theta} - \frac{2d}{\sin^2\theta}\cdot\cos\theta$$

From $\partial p/\partial\theta = 0$, we get

$$\frac{d\cdot\cos^2\theta}{\sin^2\theta}\cdot\frac{1}{\cos^2\theta} = \frac{2d}{\sin^2\theta}\cdot\cos\theta.$$

Since $\sin\theta \neq 0$ and $\cos\theta \neq 0$, canceling gives

$$1 = 2\cos\theta$$

so

$$\cos\theta = \frac{1}{2}.$$

Since $0 < \theta < \dfrac{\pi}{2}$, we get $\theta = \dfrac{\pi}{3}$.

Substituting into the equation $\partial p / \partial d = 0$ and solving for d gives:

$$\frac{-50}{d^2} - \frac{1}{\sqrt{3}} + \frac{2}{\sqrt{3}/2} = 0$$

which leads to

$$d = \sqrt{\frac{50}{\sqrt{3}}} \approx 5.37 \text{m}.$$

Then

$$w = \frac{50}{d} - \frac{d}{\tan \theta} \approx \frac{50}{5.37} - \frac{5.37}{\sqrt{3}} \approx 6.21 \text{ m}.$$

When $\theta = \pi/3$, $w \approx 6.21$ m and $d \approx 5.37$ m, we have $p \approx 18.61$ m.

Since there is only one critical point, and since p increases without limit as d or θ shrink to zero, the critical point must give the global minimum for p.

21. (a)

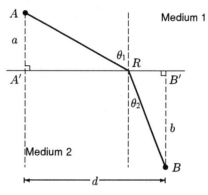

Figure 14.11

See Figure 14.11. The time to travel from A to B is given by

$$T(\theta_1, \theta_2) = \frac{AR}{v_1} + \frac{RB}{v_2} = \frac{a}{v_1 \cos \theta_1} + \frac{b}{v_2 \cos \theta_2}.$$

(b) The distance $d = A'B' = A'R + RB'$. Hence

$$d = a \tan \theta_1 + b \tan \theta_2.$$

(c) We imagine the following extreme case: the light ray first travels through medium 1 to a point R on the boundary far to the left of A', then through medium 2 towards B. The distance traveled this way is very large, hence the travel time is large as well. Similarly, if R is far to the right of B', the travel time will be large. Therefore values of θ_1 near $-\pi/2$ or $\pi/2$ increase the time, T.

(d) The constrained optimization problem is: minimize $T(\theta_1, \theta_2)$ subject to $g(\theta_1, \theta_2) = a \tan \theta_1 + b \tan \theta_2 = d$. According to the method of Lagrange multipliers, the minimum point should be among those satisfying grad $T = \lambda$ grad g as well as the constraint. We have

$$\text{grad}\, T(\theta_1, \theta_2) = \frac{a}{v_1} \frac{\sin \theta_1}{\cos^2 \theta_1} \vec{i} + \frac{b}{v_2} \frac{\sin \theta_2}{\cos^2 \theta_2} \vec{j}$$

and
$$\operatorname{grad} g(\theta_1, \theta_2) = a\frac{1}{\cos^2 \theta_1}\vec{i} + b\frac{1}{\cos^2 \theta_2}\vec{j}\,.$$

The condition $\operatorname{grad} T = \lambda \operatorname{grad} g$ becomes

$$\frac{a}{v_1}\frac{\sin \theta_1}{\cos^2 \theta_1} = \lambda a\frac{1}{\cos^2 \theta_1} \quad \text{and} \quad \frac{b}{v_2}\frac{\sin \theta_2}{\cos^2 \theta_2} = \lambda b\frac{1}{\cos^2 \theta_2}.$$

Eliminating λ we are left with
$$\frac{\sin \theta_1}{v_1} = \frac{\sin \theta_2}{v_2}$$

or
$$\frac{\sin \theta_1}{\sin \theta_2} = \frac{v_1}{v_2},$$

which is Snell's law. The argument in part (c) shows that the critical point corresponding to θ_1, θ_2 satisfying Snell's law is indeed a minimum.

CHAPTER FIFTEEN

Solutions for Section 15.1

1. Mark the values of the function on the plane, as shown in Figure 15.1, so that you can guess respectively at the smallest and largest values the function takes on each small rectangle.

$$\text{Lower sum} = \sum f(x_i, y_i)\Delta x \Delta y$$
$$= 4\Delta x \Delta y + 6\Delta x \Delta y + 3\Delta x \Delta y + 4\Delta x \Delta y$$
$$= 17\Delta x \Delta y$$
$$= 17(0.1)(0.2)$$
$$= 0.34.$$

$$\text{Upper sum} = \sum f(x_i, y_i)\Delta x \Delta y$$
$$= 7\Delta x \Delta y + 10\Delta x \Delta y + 6\Delta x \Delta y + 8\Delta x \Delta y$$
$$= 31\Delta x \Delta y$$
$$= 31(0.1)(0.2)$$
$$= 0.62.$$

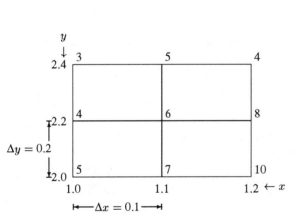

Figure 15.1

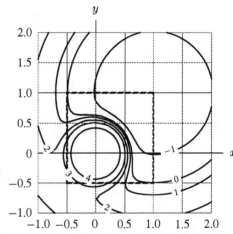

Figure 15.2

5. The total area of the square R is $(1.5)(1.5) = 2.25$. See Figure 15.2. On a disk of radius ≈ 0.5 the function has a value of 3 or more, giving a total contribution to the integral of at least $(3) \cdot (\pi \cdot 0.5^2) \approx 2.3$. On less than half of the rest of the square the function has a value between -2 and 0, giving a contribution to the integral of between $(1/2 \cdot 2.25)(-2) = -2.25$ and 0. Since the positive contribution to the integral is therefore greater in magnitude than the negative contribution, $\int_R f \, dA$ is positive.

9. Let R be the region $0 \le x \le 60, \quad 0 \le y \le 8$. Then

$$\text{Volume} = \int_R w(x, y) \, dA$$

Lower estimate:
$$10 \cdot 2(1+4+8+10+10+8+0+3+4+6+6+4+0+1+2+3+3+2+0+0+1+1+1+1) = 1580.$$
Upper estimate:
$$10 \cdot 2(8+13+16+17+17+16+4+8+10+11+11+10+3+4+6+7+7+6+1+2+3+4+4+3) = 3820.$$
The average of the two estimates is 2700 cubic feet.

13. The region D is symmetric both with respect to x and y axes, while in the interior of R, $x > 0$ and in the interior of B, $y < 0$.

   (a) The function being integrated is $f(x,y) = 1$, which is positive everywhere. Thus, its integral over any region is positive.

   (b) For the same reason as in part (a), the integral is positive.

   (c) The function being integrated is $f(x,y) = 5x$. Since $x > 0$ in R, f is positive in R and thus the integral is positive.

   (d) The function being integrated is $f(x,y) = 5x$, which is an odd function in x. Since B is symmetric with respect to x, the contributions to the integral will cancel out, as $f(x,y) = -f(-x,y)$. Thus, the integral is zero.

   (e) As in part (d), the function is an odd function in x. Since D is also symmetric with respect to x, the integral of the function over the area D is zero.

   (f) The function being integrated, $f(x,y) = y^3 + y^5$, is an odd function in y while D is symmetric with respect to y. Then, by symmetry, the positive and negative contributions of f are equal and thus its integral is zero.

   (g) In a region such as R in which $y < 0$, the quantity $y^3 + y^5$ is less than zero. Thus, its integral is negative.

   (h) Using the same reasoning as in part (f), we determine that the integral over B is zero.

   (i) The function being integrated, $f(x,y) = y - y^3$ is always negative in the region B since in that region $-1 < y < 0$. Thus, the integral is negative.

   (j) The function being integrated, $f(x,y) = y - y^3$, is an odd function in y while D is symmetric with respect to y. By symmetry, the integral is zero.

   (k) As in part (f), the function being integrated is odd with respect to y in the region D. Thus, its integral is zero.

   (l) In the region D, y has range $|y| < 1$. Thus, $-\pi/2 < y < \pi/2$. Thus, $\cos y$ is always positive in the region D and thus its integral is positive.

   (m) The function $f(x,y) = e^x$ is positive for any value of x. Thus, its integral is always positive for any region, such as D, with non-zero area.

   (n) Looking at the contributions to the integral of the function $f(x,y) = xe^x$, we can see that any contribution made by the point (x,y), where $x > 0$, is greater than the corresponding contribution made by $(-x,y)$, since $e^x > 1 > e^{-x}$ for $x > 0$. Thus, the integral of f in the region D is positive.

   (o) The function $f(x,y) = xy^2$ is odd with respect to x and thus, using the same reasoning as in part (d), has integral zero in the region D.

   (p) The function $f(x,y)$ is odd with respect to x and thus, using the same reasoning as in part (d), has integral zero in region B, which is symmetric with respect to x.

Solutions for Section 15.2

1.

$$\int_R \sqrt{x+y} \, dA = \int_0^2 \int_0^1 \sqrt{x+y} \, dx \, dy$$

$$= \int_0^2 \frac{2}{3}(x+y)^{\frac{3}{2}} \Big|_0^1 \, dy$$

$$= \frac{2}{3} \int_0^2 ((1+y)^{\frac{3}{2}} - y^{\frac{3}{2}})\, dy$$

$$= \frac{2}{3} \cdot \frac{2}{5} [(1+y)^{\frac{5}{2}} - y^{\frac{5}{2}}]\Big|_0^2$$

$$= \frac{4}{15} ((3^{\frac{5}{2}} - 2^{\frac{5}{2}}) - (1-0))$$

$$= \frac{4}{15} (9\sqrt{3} - 4\sqrt{2} - 1) = 2.38176$$

5. $\int_1^4 \int_1^2 f\, dy\, dx$ or $\int_1^2 \int_1^4 f\, dx\, dy$

9. $\int_1^3 \int_0^4 e^{x+y}\, dx dy = \int_1^3 e^x e^y \Big|_0^4\, dx = \int_1^3 e^x (e^4 - 1)\, dx = (e^4 - 1)(e^2 - 1)e$
    See Figure 15.3.

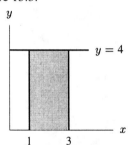

Figure 15.3

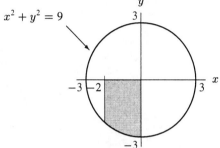

Figure 15.4

13. See Figure 15.4.

$$\int_{-2}^0 \int_{-\sqrt{9-x^2}}^0 2xy\, dy dx = \int_{-2}^0 x y^2 \Big|_{-\sqrt{9-x^2}}^0\, dx$$

$$= -\int_{-2}^0 x(9 - x^2)\, dx$$

$$= \int_{-2}^0 (x^3 - 9x)\, dx$$

$$= \left(\frac{x^4}{4} - \frac{9}{2}x^2 \right) \Big|_{-2}^0$$

$$= -4 + 18 = 14$$

17. As given, the region of integration is as shown in Figure 15.5.

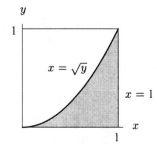

Figure 15.5

Reversing the limits gives

$$\int_0^1 \int_0^{x^2} \sqrt{2+x^3}\, dy dx = \int_0^1 (y\sqrt{2+x^3}\Big|_0^{x^2})\, dx$$

$$= \int_0^1 x^2\sqrt{2+x^3}\, dx$$

$$= \frac{2}{9}(2+x^3)^{\frac{3}{2}}\Big|_0^1$$

$$= \frac{2}{9}(3\sqrt{3} - 2\sqrt{2}).$$

21. The solid is shown in Figure 15.6, and the base of the integral is the triangle as shown in Figure 15.7.

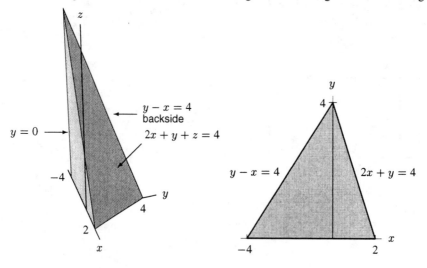

<table>
<tr><td>Figure 15.6</td><td>Figure 15.7</td></tr>
</table>

Thus, the volume of the solid is

$$V = \int_R z\, dA$$

$$= \int_R (4 - 2x - y)\, dA$$

$$= \int_0^4 \int_{y-4}^{(4-y)/2} (4 - 2x - y)\, dx\, dy.$$

25. The region bounded by the x-axis and the graph of $y = x - x^2$ is shown in Figure 15.8. The area of this region is

$$A = \int_0^1 (x - x^2)dx = (\frac{x^2}{2} - \frac{x^3}{3})\Big|_0^1$$

$$= \frac{1}{2} - \frac{1}{3} = \frac{1}{6}.$$

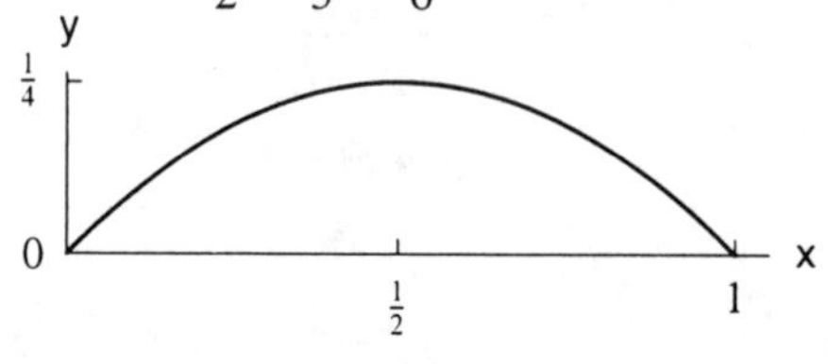

Figure 15.8

So the average distance to the x-axis for points in the region is

$$\text{Average distance} = \frac{\int_R y \, dA}{\text{area}(R)}$$

$$\int_R y \, dA = \int_0^1 \left(\int_0^{x-x^2} y \, dy \right) dx$$

$$= \int_0^1 \left(\frac{x^2}{2} - x^3 + \frac{x^4}{2} \right) dx = \frac{1}{6} - \frac{1}{4} + \frac{1}{10} = \frac{1}{60}.$$

Therefore the average distance is $\frac{1/60}{1/6} = 1/10$.

29. We want to find the average value of $|x - y|$, over the square $0 \le x \le 1, 0 \le y \le 1$:

$$\text{Average distance between gates} = \int_0^1 \int_0^1 |x - y| \, dy \, dx.$$

Let's fix x, with $0 \le x \le 1$. Then $|x - y| = \begin{cases} y - x & \text{for } y \ge x \\ x - y & \text{for } y \le x \end{cases}$. Therefore

$$\int_0^1 |x - y| \, dy = \int_0^x (x - y) \, dy + \int_x^1 (y - x) \, dy$$

$$= \left(xy - \frac{y^2}{2} \right) \Big|_0^x + \left(\frac{y^2}{2} - xy \right) \Big|_x^1 = x^2 - \frac{x^2}{2} + \frac{1}{2} - x - \frac{x^2}{2} + x^2$$

$$= x^2 - x + \frac{1}{2}.$$

So,

$$\text{Average distance between gates} = \int_0^1 \int_0^1 |x - y| \, dy \, dx$$

$$= \int_0^1 \left(\int_0^1 |x - y| \, dy \right) dx = \int_0^1 \left(x^2 - x + \frac{1}{2} \right) dx$$

$$= \frac{x^3}{3} - \frac{x^2}{2} + \frac{1}{2} x \Big|_0^1 = \frac{1}{3}.$$

Solutions for Section 15.3

1.

$$\int_W f \, dV = \int_0^2 \int_{-1}^1 \int_2^3 (x^2 + 5y^2 - z) \, dz \, dy \, dx$$

$$= \int_0^2 \int_{-1}^1 (x^2 z + 5y^2 z - \frac{1}{2} z^2) \Big|_2^3 \, dy \, dx$$

$$= \int_0^2 \int_{-1}^1 (x^2 + 5y^2 - \frac{5}{2}) \, dy \, dx$$

5. The region of integration for this integral is shown in Figure 15.7.

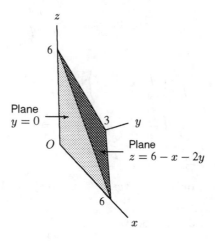

Figure 15.7

This is a three-sided pyramid whose base is the xy-plane and whose three sides are the vertical planes $x = 0, y = 0$ and slanted plane $z = 6 - x - y$.

9. The limits do not make sense since the limits for the middle integral involve two variables.

13. The region of integration is shown in Figure 15.8, and the mass of the given solid is given by

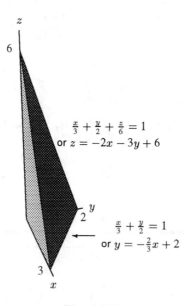

Figure 15.8

$$\text{mass} = \int_R \delta \, dV$$

$$= \int_0^3 \int_0^{-\frac{2}{3}x+2} \int_0^{-2x-3y+6} (x+y)\,dz\,dy\,dx$$

$$= \int_0^3 \int_0^{-\frac{2}{3}x+2} (x+y)z \Big|_0^{-2x-3y+6} dy\,dx$$

$$= \int_0^3 \int_0^{-\frac{2}{3}x+2} (x+y)(-2x-3y+6)\,dy\,dx$$

$$= \int_0^3 \int_0^{-\frac{2}{3}x+2} (-2x^2 - 3y^2 - 5xy + 6x + 6y)\,dy\,dx$$

$$= \int_0^3 \left(-2x^2y - y^3 - \frac{5}{2}xy^2 + 6xy + 3y^2\right) \Big|_0^{-\frac{2}{3}x+2} dx$$

$$= \int_0^3 \left(\frac{14}{27}x^3 - \frac{8}{3}x^2 + 2x + 4\right) dx$$

$$= \left(\frac{7}{54}x^4 - \frac{8}{9}x^3 + x^2 + 4x\right) \Big|_0^3$$

$$= \frac{7}{54}\cdot 3^4 - \frac{8}{9}\cdot 3^3 + 3^2 + 12 = \frac{21}{2} - 3 = \frac{15}{2}.$$

17. The mass m is given by

$$m = \int_W 1\,dV = \int_0^1 \int_0^{(1-x)/2} \int_0^{(1-x-2y)/3} 1\,dz\,dy\,dx$$

$$= \int_0^1 \int_0^{(1-x)/2} \frac{1-x-2y}{3}\,dy\,dx$$

$$= \frac{1}{3} \int_0^1 (y - xy - y^2) \Big|_0^{(1-x)/2} dx$$

$$= \frac{1}{3} \left(\int_0^1 \frac{1-x}{2} - x\frac{1-x}{2} - \left(\frac{1-x}{2}\right)^2\right) dx = 1/36 \text{ gm.}$$

Then the coordinates of the center of mass are given by

$$\bar{x} = 36 \int_W x\,dV = 36 \int_0^1 \int_0^{(1-x)/2} \int_0^{(1-x-2y)/3} x\,dz\,dy\,dx = 1/4 \text{ cm.}$$

and

$$\bar{y} = 36 \int_W y\,dV = 36 \int_0^1 \int_0^{(1-x)/2} \int_0^{(1-x-2y)/3} y\,dz\,dy\,dx = 1/8 \text{ cm.}$$

and

$$\bar{z} = 36 \int_W z\,dV = 36 \int_0^1 \int_0^{(1-x)/2} \int_0^{(1-x-2y)/3} z\,dz\,dy\,dx = 1/12 \text{ cm.}$$

Solutions for Section 15.4

1. We enclose the area represented by the integral in a square of area 1. Here is a table of values of one run of the method for different numbers of points.

TABLE 15.1

N	10^3	10^4	10^5	10^6	10^7
N_R/N	0.78400	0.78730	0.78544	0.78536	0.78543

The integral is the area under the graph of $\sqrt{1-x^2}$ for $0 \leq x \leq 1$, which is a quarter-circle of radius 1, so the integral is $\pi/4$. Since $\pi/4 \approx 0.7853981635$, we see from the table that, for large N, our results are comparable to about the fourth place.

5. In this problem, we see that the graph of $z = x\sin(y)$ over the region $0 \leq x \leq 2$, $0 \leq y \leq \pi$ lies between 0 and 2. So we'll have to generate random z and x between 0 and 2, and random y between 0 and π. We enclose the volume represented by the integral in a box of volume $2 \cdot \pi \cdot 2 = 4\pi$. Here is a table of values of one run of the method for different numbers of points:

TABLE 15.2

N	10^3	10^4	10^5	10^6	10^7
N_R/N	0.305	0.31299	0.31785	0.31796	0.31831

Since

$$\frac{N_R}{N} \approx \frac{\int_R f \, dA}{\text{Vol}(C)}$$

our estimate for the integral is

$$\int_R f \, dA \approx \text{Vol}(C)\frac{N_R}{N} = 4\pi\frac{N_R}{N}.$$

Multiplying the last estimate (0.31831) by 4π gives 4.00000143, which is very close to the exact value of 4. The exact value is found by calculating

$$\int_0^\pi \int_0^2 x\sin y \, dx \, dy = \int_0^\pi (\sin y)\frac{x^2}{2}\Big|_0^2 \, dy = 2\int_0^\pi \sin y \, dy = -2\cos y\Big|_0^\pi = 4.$$

9. The area of the region is 2π, so we need to compute the average $a = \frac{1}{N}\sum_{i=1}^N f(x_i, y_i)$ and multiply by 2π. Here is a table of values of one run of the method for different numbers of points:

TABLE 15.3

N	10^3	10^4	10^5	10^6	10^7
$2\pi a$	3.79003	3.81567	3.760207	3.9900132	4.0000012

The values seem to be getting closer to 4 (which is the exact answer).

Solutions for Section 15.5

1.

$$\int_{\pi/4}^{3\pi/4} \int_0^2 f \, r \, dr \, d\theta$$

5.

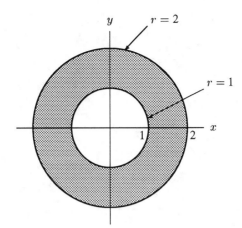

Figure 15.9

9.

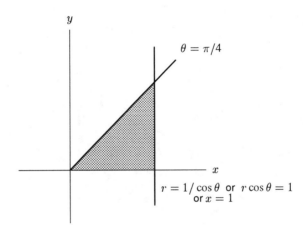

Figure 15.10

13.   The region is pictured in Figure 15.11.

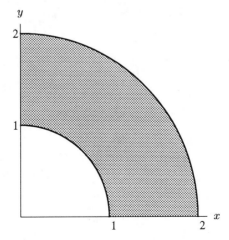

Figure 15.11

By using polar coordinates, we get

$$\int_R (x^2 - y^2)\,dA = \int_0^{\pi/2} \int_1^2 r^2(\cos^2\theta - \sin^2\theta)r\,dr\,d\theta = \int_0^{\pi/2} (\cos^2\theta - \sin^2\theta)\cdot\frac{1}{4}r^4\Big|_1^2 d\theta$$

$$= \frac{15}{4}\int_0^{\pi/2} (\cos^2\theta - \sin^2\theta)\,d\theta$$

$$= \frac{15}{4}\int_0^{\pi/2} \cos 2\theta\,d\theta$$

$$= \frac{15}{4}\cdot\frac{1}{2}\sin 2\theta\Big|_0^{\pi/2}$$

$$= 0$$

17.   From the given limits, the region of integration is in Figure 15.12.

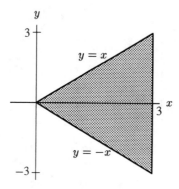

Figure 15.12

In polar coordinates, $-\pi/4 \le \theta \le \pi/4$. Also, $3 = x = r\cos\theta$. Hence, $0 \le r \le 3/\cos\theta$. The integral becomes

$$\int_{-\pi/4}^{\pi/4} \int_0^{3/\cos\theta} \frac{r\cos\theta}{(r\sin\theta)^2} r\,dr\,d\theta$$

$$= \int_{-\pi/4}^{\pi/4} \frac{\cos\theta}{\sin^2\theta}\left(\frac{3}{\cos\theta}\right) d\theta$$

$$= 3\int_{-\pi/4}^{\pi/4} \frac{d\theta}{\sin^2\theta}$$

$$= -3\frac{\cos\theta}{\sin\theta}\Big|_{-\pi/4}^{\pi/4}$$

$$= -3\cdot(-1 - (-1)) = 6.$$

21. (a)

$$\text{Total Population} = \int_{\pi/2}^{3\pi/2} \int_1^4 \delta(r, \theta)\, r\, dr\, d\theta.$$

(b) We know that $\delta(r, \theta)$ decreases as r increases, so that eliminates (iii). We also know that $\delta(r, \theta)$ decreases as the x-coordinate decreases, but $x = r\cos\theta$. With a fixed r, x is proportional to $\cos\theta$. So as the x-coordinate decreases, $\cos\theta$ decreases and (i) $\delta(r, \theta) = (4 - r)(2 + \cos\theta)$ best describes this situation.

(c)

$$\int_{\pi/2}^{3\pi/2} \int_1^4 (4 - r)(2 + \cos\theta)\, r\, dr\, d\theta = \int_{\pi/2}^{3\pi/2} (2 + \cos\theta)\left(2r^2 - \frac{1}{3}r^3\right)\Big|_1^4 d\theta$$

$$= 9\int_{\pi/2}^{3\pi/2} (2 + \cos\theta)\, d\theta$$

$$= 9\left[2\theta + \sin\theta\right]_{\pi/2}^{3\pi/2}$$

$$= 18(\pi - 1)$$

$$\approx 39$$

Thus, the population is around $39,000$.

Solutions for Section 15.6

1.

$$\int_W f\, dV = \int_{-1}^1 \int_{\pi/4}^{3\pi/4} \int_0^4 (r^2 + z^2)\, r\, dr\, d\theta\, dz$$

$$= \int_{-1}^1 \int_{\pi/4}^{3\pi/4} (64 + 8z^2)\, d\theta\, dz$$

$$= \int_{-1}^1 \frac{\pi}{2}(64 + 8z^2)\, dz$$

$$= 64\pi + \frac{8}{3}\pi$$

$$= \frac{200}{3}\pi$$

5. Using cylindrical coordinates, we get:

$$\int_0^1 \int_0^{2\pi} \int_0^4 \delta \cdot r\, dr\, d\theta\, dz$$

9. Using Cartesian coordinates, we get:

$$\int_0^3 \int_0^1 \int_0^5 \delta\, dz\, dy\, dx$$

13. (a) The angle ϕ takes on values in the range $0 \leq \phi \leq \pi$. Thus, $\sin \phi$ is nonnegative everywhere in W_1, and so its integral is positive.

(b) The function ϕ is symmetric across the xy plane, such that for any point (x, y, z) in W_1, with $z \neq 0$, the point $(x, y, -z)$ has a $\cos \phi$ value with the same magnitude but opposite sign of the $\cos \phi$ value for (x, y, z). Furthermore, if $z = 0$, then (x, y, z) has a $\cos \phi$ value of 0. Thus, with $\cos \phi$ positive on the top half of the sphere and negative on the bottom half, the integral will cancel out and be equal to zero.

17. Using spherical coordinates:

$$
\begin{aligned}
M &= \int_0^\pi \int_0^{2\pi} \int_0^3 (3 - \rho)\rho^2 \sin\phi \, d\rho \, d\theta \, d\phi \\
&= \int_0^\pi \int_0^{2\pi} \left[\rho^3 - \frac{\rho^4}{4} \right]_0^3 \sin\phi \, d\theta \, d\phi \\
&= \frac{27}{4} \int_0^\pi \int_0^{2\pi} \sin\phi \, d\theta \, d\phi \\
&= \frac{27}{4} \cdot 2\pi \cdot [-\cos\phi]_0^\pi \\
&= \frac{27}{4} \cdot 2\pi \cdot [-(-1) + 1] \\
&= 27\pi.
\end{aligned}
$$

21. The total volume of the cone is $\frac{1}{3}\pi r^2 h = \frac{1}{3}\pi \cdot 1^2 \cdot 1 = \frac{1}{3}\pi$, so the total mass is $\frac{1}{3}\pi$ (since the density is always 1). The center of mass z-coordinate is given by

$$
\bar{z} = \frac{3}{\pi} \int_C z \, dV
$$

Using cylindrical coordinates to evaluate this integral gives

$$
\begin{aligned}
\bar{z} &= \frac{3}{\pi} \int_0^{2\pi} \int_0^1 \int_0^z zr \, dr \, dz \, d\theta \\
&= \frac{3}{\pi} \int_0^{2\pi} \int_0^1 \frac{z^3}{2} \, dz \, d\theta \\
&= \frac{3}{\pi} \int_0^{2\pi} \frac{1}{8} \, d\theta = \frac{3}{4}
\end{aligned}
$$

25. The sum of the three moments of inertia I for the ball B will be

$$
\begin{aligned}
3I &= \frac{3}{4\pi a^3} \int_B (y^2 + z^2) \, dV + \frac{3}{4\pi a^3} \int_B (x^2 + z^2) \, dV + \frac{3}{4\pi a^3} \int_B (x^2 + y^2) \, dV \\
&= \frac{3}{4\pi a^3} \int_B (2x^2 + 2y^2 + 2z^2) \, dV,
\end{aligned}
$$

which, in spherical coordinates is

$$
\frac{3}{2\pi a^3} \int_B (x^2 + y^2 + z^2) \, dV = \frac{3}{2\pi a^3} \int_0^a \int_0^\pi \int_0^{2\pi} \rho^2 \cdot \rho^2 \sin(\phi) \, d\theta \, d\phi \, d\rho
$$

$$= \frac{3}{a^3} \int_0^a \int_0^\pi \rho^4 \sin(\phi) \, d\phi \, d\rho$$

$$= \frac{6}{a^3} \int_0^a \rho^4 \, d\rho = \frac{6}{5}a^2.$$

Thus $3I = \frac{6}{5}a^2$, so $I = \frac{2}{5}a^2$.

Solutions for Section 15.7

1. (a)

$$\int_0^1 \int_{1/3}^1 \frac{2}{3}(x + 2y) \, dx \, dy = \int_0^1 \frac{2}{3}(\frac{1}{2}x^2 + 2xy)\Big|_{1/3}^1 \, dy$$

$$= \int_0^1 \frac{2}{3}\left[(\frac{1}{2} + 2y) - (\frac{1}{18} + \frac{2}{3}y)\right] \, dy$$

$$= \frac{2}{3} \int_0^1 (\frac{4}{9} + \frac{4}{3}y) \, dy$$

$$= \frac{2}{3} \left(\frac{4}{9}y + \frac{2}{3}y^2\right)\Big|_0^1$$

$$= \frac{2}{3} \left(\frac{10}{9}\right) = \frac{20}{27}.$$

(b) It is easier to calculate the probability that $x < (1/3) + y$ does not happen, that is, the probability that $x \geq (1/3) + y$, and subtract it from 1. The probability that $x \geq (1/3) + y$ is

$$\int_{1/3}^1 \int_0^{x-(1/3)} \frac{2}{3}(x + 2y) \, dy \, dx = \int_{1/3}^1 \frac{2}{3}(xy + y^2)\Big|_0^{x-(1/3)} \, dx$$

$$= \frac{2}{3} \int_{1/3}^1 (x(x - \frac{1}{3}) + (x - \frac{1}{3})^2) \, dx$$

$$= \frac{2}{3} \int_{1/3}^1 (2x^2 - x + \frac{1}{9}) \, dx$$

$$= \frac{2}{3}(\frac{2}{3}x^3 - \frac{1}{2}x^2 + \frac{1}{9}x)\Big|_{1/3}^1$$

$$= \frac{2}{3}\left[(\frac{2}{3} - \frac{1}{2} + \frac{1}{9}) - (\frac{2}{81} - \frac{1}{18} + \frac{1}{27})\right]$$

$$= 44/243.$$

Thus, the probability that $x < (1/3) + y$ is $1 - (44/243) = 199/243$.

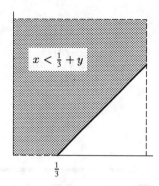

$$x < \tfrac{1}{3} + y$$

$$\tfrac{1}{3}$$

Figure 15.13

5. Since

$$\sum_x \sum_y f(x,y)\,\Delta x\,\Delta y \approx \int_R f(x,y)\,dx\,dy$$

and since x never exceeds 1, and we can assume that no one lives to be over 100, so y does not exceed 100, we have

$$\text{Fraction of policies} = \int_R f(x,y)\,dx\,dy = \int_{65}^{100}\int_{0.8}^{1} f(x,y)\,dx\,dy,$$

where R is the rectangle: $0.8 \le x \le 1,\, 65 \le y \le 100$.

9. (a)

$$\int_{\theta=0}^{\frac{\pi}{6}}\int_{r=\frac{1}{\cos\theta}}^{4} p(r,\theta)r\,dr\,d\theta$$

(b)

$$\int_{\theta=\frac{\pi}{6}}^{\frac{\pi}{6}+\frac{\pi}{12}}\int_{r=\frac{1}{\cos\theta}}^{4} p(r,\theta)r\,dr\,d\theta + \int_{\theta=\frac{\pi}{6}+\frac{\pi}{12}}^{\frac{2\pi}{6}}\int_{r=\frac{1}{\sin\theta}}^{4} p(r,\theta)r\,dr\,d\theta$$

Solutions for Section 15.8

1. Given $T = \{(s,t)\,|\, 0 \le s \le 3,\, 0 \le t \le 2\}$ and

$$\begin{cases} x = 2s - 3t \\ y = \ s - 2t \end{cases}$$

The shaded area in Figure 15.14 is the corresponding region R in the xy-plane.
    Since

$$\frac{\partial(x,y)}{\partial(s,t)} = \begin{vmatrix} \frac{\partial x}{\partial s} & \frac{\partial x}{\partial t} \\ \frac{\partial y}{\partial s} & \frac{\partial y}{\partial t} \end{vmatrix} = \begin{vmatrix} 2 & -3 \\ 1 & -2 \end{vmatrix} = -1,$$

$$\left|\frac{\partial(x,y)}{\partial(s,t)}\right| = 1.$$

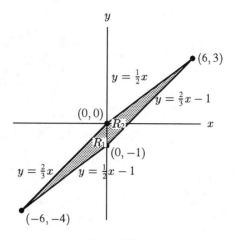

Figure 15.14

Thus we get

$$\int_T \left| \frac{\partial(x,y)}{\partial(s,t)} \right| ds\, dt = \int_0^3 ds \int_0^2 dt = 6.$$

Since

$$\int_R dx\, dy = \int_{R_1} dx\, dy + \int_{R_2} dx\, dy = \int_{-6}^0 dx \int_{\frac{1}{2}x-1}^{\frac{2}{3}x} dy + \int_0^6 dx \int_{\frac{2}{3}x-1}^{\frac{1}{2}x} dy$$

$$= \int_{-6}^0 \left(\frac{1}{6}x + 1\right) dx + \int_0^6 \left(\frac{-1}{6}x + 1\right) dx = 3 + 3 = 6,$$

thus

$$\int_R dx\, dy = \int_T \left| \frac{\partial(x,y)}{\partial(s,t)} \right| ds\, dt.$$

5. Given

$$\begin{cases} x = 2s + t \\ y = s - t, \end{cases}$$

we have

$$\frac{\partial(x,y)}{\partial(s,t)} = \begin{vmatrix} \frac{\partial x}{\partial s} & \frac{\partial x}{\partial t} \\ \frac{\partial y}{\partial s} & \frac{\partial y}{\partial t} \end{vmatrix} = \begin{vmatrix} 2 & 1 \\ 1 & -1 \end{vmatrix} = -3,$$

hence

$$\left| \frac{\partial(x,y)}{\partial(s,t)} \right| = 3.$$

We get

$$\int_R (x+y)\, dA = \int_T 3s \left| \frac{\partial(x,y)}{\partial(s,t)} \right| ds dt = \int_T (3s)(3)\, ds\, dt = 9 \int_T s\, ds\, dt,$$

where T is the region in the st-plane corresponding to R.

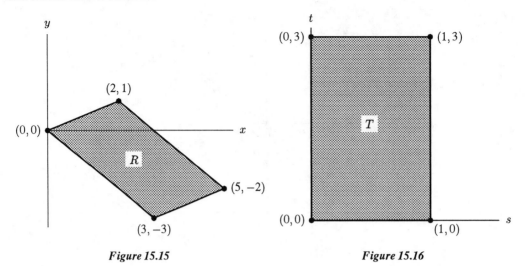

Figure 15.15   Figure 15.16

Now, we need to find T.

As

$$\begin{cases} x = 2s + t \\ y = s - t \end{cases} \quad \text{or} \quad \begin{cases} s = \frac{1}{3}(x + y) \\ t = \frac{1}{3}(x - 2y), \end{cases}$$

so from the above transformation and Figure 15.15, T is the shaded area in Figure 15.16. Therefore

$$\int_R (x + y) \, dA = 9 \int_0^1 s \, ds \int_0^3 dt = (27)(\frac{1}{2}) = 13.5.$$

Solutions for Chapter 15 Review

1. First we will subdivide the area into 70 squares, as shown. We will find the upper and lower bounds for the total rainfall, and then take the average of the two. We do this by finding the highest amount of rainfall in that subdivision, multiplying it by the area, and then adding up all the contributions from each subdivisions, and then doing the same for the lowest amount of rainfall.

   For an upper bound on the rainfall, going left to right, top to bottom, we get: $[(80+40+40+40+0+0+0)+(80+80+80+40+40+0+0)+(80+80+80+40+40+40+80)+(80+80+80+80+40+40+80)+(12+40+40+40+80+80+40)+(20+12+20+80+80+80+40)+(20+20+20+40+80+40+0)+(40+40+40+20+0+40+40)+(0+0+40+20+40+40+0)+(0+0+40+40+40+0+0)] = 2804$ in. This gives a volume of $(2804 \text{ in})(0.00001578 \text{ miles/in})(500 \text{ miles})(500 \text{ miles})$. This equals about 11061 cubic miles for the upper bound. For the low rainfall we get: $[(40+20+20+40+0+0+0)+(80+40+40+40+40+0+0)+(12+80+80+40+40+40+40)+(4+4+80+80+40+40+40)+(4+4+4+40+40+40+40)+(12+4+4+40+80+40+40)+(20+4+4+40+40+40+0)+(40+40+12+20+0+40+40)+(0+0+12+12+40+40+0)+(0+0+12+12+40+0+0)](0.00001578 \text{ miles})(500 \text{ miles})(500 \text{ miles})$. This equals about 7417 cubic miles for the lower bound. Taking the average of these two, we get $(11061 + 7417)/2 = 9239$ cubic miles of rain over a year.

5.

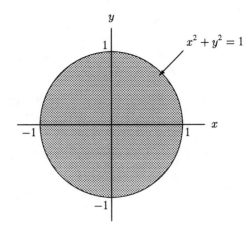

Figure 15.17

9.

$$\int_0^{10} \int_0^{0.1} xe^{xy} \, dy \, dx = \int_0^{10} e^{xy}\big|_0^{0.1} \, dx$$

$$= \int_0^{10} (e^{0.1x} - e^0) \, dx$$

$$= \left(\left(\frac{e^{0.1x}}{0.1} \right) - x \right) \Big|_0^{10}$$

$$= (10e^1 - 10 - 10e^0)$$

$$= 10e - 20 = 10(e - 2).$$

13.

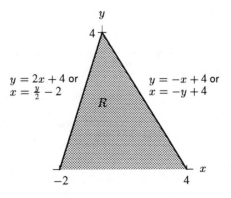

Figure 15.18

Integrating with respect to x first we get

$$\int_0^4 \int_{\frac{y}{2}-2}^{-y+4} f(x,y) \, dx \, dy$$

Integrating with respect to y first we get

$$\int_{-2}^{0} \int_{0}^{2x+4} f(x,y)\, dy\, dx + \int_{0}^{4} \int_{0}^{-x+4} f(x,y)\, dy\, dx.$$

17. The region is a hemisphere $0 \le x^2 + y^2 + z^2 \le 3^2$, $z \ge 0$, so spherical coordinates are appropriate. Recall the conversion formula $x = \rho \sin \phi \cos \theta$. Then the integral in spherical coordinates becomes

$$\int_{0}^{2\pi} \int_{0}^{\pi/2} \int_{0}^{3} (\rho \sin \phi \cos \theta)^2 \rho^2 \sin \phi\, d\rho\, d\phi\, d\theta$$

$$= \int_{0}^{2\pi} \int_{0}^{\pi/2} \int_{0}^{3} \rho^4 \sin^3 \phi \cos^2 \theta\, d\rho\, d\phi\, d\theta$$

$$= \int_{0}^{2\pi} \int_{0}^{\pi/2} \frac{243}{5} \sin^3 \phi \cos^2 \theta\, d\phi\, d\theta$$

$$= \frac{243}{5} \int_{0}^{2\pi} \int_{0}^{\pi/2} \cos^2 \theta \cdot \sin \phi (1 - \cos^2 \phi)\, d\phi\, d\theta$$

$$= \frac{243}{5} \int_{0}^{2\pi} \cos^2 \theta \left[-\cos \phi + \frac{1}{3} \cos^3 \phi \right]_{0}^{\frac{\pi}{2}} d\theta$$

$$= \frac{243}{5} \int_{0}^{2\pi} \cos^2 \theta [-(-1) + \frac{1}{3}(-1)]\, d\theta$$

$$= \frac{243}{5} \cdot \frac{2}{3} \int_{0}^{2\pi} \frac{1 + \cos 2\theta}{2}\, d\theta$$

$$= \frac{81}{5} (\theta + \frac{1}{2} \sin 2\theta) \Big|_{0}^{2\pi}$$

$$= \frac{81}{5} (2\pi + 0)$$

$$= \frac{162\pi}{5}$$

21. Let the lower left part of the forest be at $(0,0)$. Then the other corners have coordinates as shown. The population density function is then given by

$$\rho(x,y) = 10 - 2y$$

The equations of the two diagonal lines are $x = -2y/5$ and $x = 6 + 2y/5$. So the total rabbit population in the forest is

$$\int_{0}^{5} \int_{-\frac{2}{5}y}^{6+\frac{2}{5}y} (10 - 2y)\, dx\, dy = \int_{0}^{5} (10 - 2y)(6 + \frac{4}{5}y)\, dy$$

$$= \int_{0}^{5} (60 - 4y - \frac{8}{5}y^2)\, dy$$

$$= (60y - 2y^2 - \frac{8}{15}y^3) \Big|_{0}^{5}$$

$$= 300 - 50 - \frac{8}{15} \cdot 125$$

$$= \frac{2750}{15} = \frac{550}{3}$$
$$\approx 183$$

25. Set up the cylinder with the base centered at the origin on the xy plane, facing up. (See Figure 15.19.)
Newton's Law of Gravitation states that the force exerted between two particles is

$$F = G\frac{m_1 m_2}{\rho^2}$$

where G is the gravitational constant, m_1 and m_2 are the masses, and ρ is the distance between the particles.
We take a small volume element, so $m_1 = m$, and $m_2 = \delta dV$. In cylindrical coordinates, if m is at $(0,0,0)$
and δdV is at (r, θ, z), (see Figure 15.19), then the distance from m to δdV is given by $\rho = \sqrt{r^2 + z^2}$ for
$r_1 \le r \le r_2$ and $0 \le z \le h$.

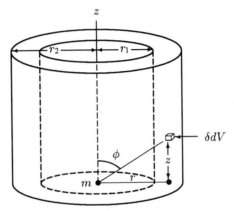

Figure 15.19

Due to the symmetry of the cylinder the sum of all the horizontal forces is zero; the net force on m is vertical.
The force acting on the particle as a result of the small piece dV makes an angle ϕ with the vertical and
therefore has vertical component

$$\text{Vertical force on particle from small piece of cylinder} = \frac{Gm\delta dV}{(\sqrt{r^2 + z^2})^2} \cdot \cos\phi = \frac{Gm\delta dV}{r^2 + z^2} \cdot \frac{z}{\sqrt{r^2 + z^2}} = \frac{Gm\delta z}{(r^2 + z^2)^{\frac{3}{2}}} dV.$$

Thus, since $dV = r\,dz\,dr\,d\theta$,

$$\text{Total force} = \int_0^{2\pi} \int_{r_1}^{r_2} \int_0^h \frac{Gm\delta z}{(r^2 + z^2)^{3/2}} r\,dz\,dr\,d\theta$$

$$= 2\pi Gm\delta \int_{r_1}^{r_2} \int_0^h \frac{zr}{(r^2 + z^2)^{3/2}}\,dr\,dz$$

$$= 2\pi Gm\delta \int_{r_1}^{r_2} \left(1 - \frac{r}{(r^2 + h^2)^{\frac{1}{2}}}\right)\,dr$$

$$= 2\pi Gm\delta (r - (r^2 + h^2)^{\frac{1}{2}}) \Big|_{r_1}^{r_2}$$

$$= 2\pi Gm\delta (r_2 - r_1 - \sqrt{r_2^2 + h^2} + \sqrt{r_1^2 + h^2}).$$

CHAPTER SIXTEEN

Solutions for Section 16.1

1. Between times $t = 0$ and $t = 1$, x goes at a constant rate from 0 to 1 and y goes at a constant rate from 1 to 0. So the particle moves in a straight line from $(0, 1)$ to $(1, 0)$. Similarly, between times $t = 1$ and $t = 2$, it goes in a straight line to $(0, -1)$, then to $(-1, 0)$, then back to $(0, 1)$. So it traces out the diamond shown in Figure 16.1.

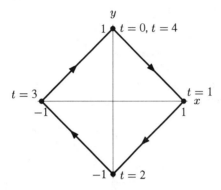

Figure 16.1

5. The particle moves clockwise: For $0 \leq t \leq \frac{\pi}{2}$, we have $x = \cos t$ decreasing and $y = -\sin t$ decreasing. Similarly, for the time intervals $\frac{\pi}{2} \leq t \leq \pi, \pi \leq t \leq \frac{3\pi}{2}$, and $\frac{3\pi}{2} \leq t \leq 2\pi$, we see that the particle moves clockwise.

9. Let $f(t) = \ln t$. Then $f'(t) = \frac{1}{t}$. The particle is moving counterclockwise when $f'(t) > 0$, that is, when $t > 0$. Any other time, when $t \leq 0$, the position is not defined.

13. One possible answer is $x = -2, y = t$.

17. The ellipse $x^2/25 + y^2/49 = 1$ can be parameterized by $x = 5\cos t$, $y = 7\sin t, 0 \leq t \leq 2\pi$.

21. (a) C_1 has center at the origin and radius 5, so $a = b = 0, k = 5$ or -5.
    (b) C_2 has center at $(0, 5)$ and radius 5, so $a = 0, b = 5, k = 5$ or -5.
    (c) C_3 has center at $(10, -10)$, so $a = 10, b = -10$. The radius of C_3 is $\sqrt{10^2 + (-10)^2} = \sqrt{200}$, so $k = \sqrt{200}$ or $k = -\sqrt{200}$.

25. The vector connecting the two points is $3\vec{i} - \vec{j} + \vec{k}$. So a possible parameterization is

$$x = 2 + 3t, \quad y = 3 - t, \quad z = -1 + t.$$

29. Let $f(x, y, z) = x^2 + y^2 - z$. Then the surface $z = x^2 + y^2$ is a level surface of f at the value 0. The gradient of f is perpendicular to the level surface.

$$\operatorname{grad} f = 2x\vec{i} + 2y\vec{j} - \vec{k} = 2\vec{i} + 4\vec{j} - \vec{k}.$$

So a possible parameterization is

$$x = 1 + 2t, \quad y = 2 + 4t, \quad z = 5 - t.$$

33. (a) Both paths are straight lines, the first passes through the point $(-1, 4, -1)$ in the direction of the vector $\vec{i} - \vec{j} + 2\vec{k}$ and the second passes through $(-7, -6, 1)$ in the direction of the vector $2\vec{i} + 2\vec{j} + \vec{k}$. The two paths are not parallel.

(b) Is there a time t when the two particles are at the same place at the same time? If so, then their coordinates will be the same, so equating coordinates we get

$$-1 + t = -7 + 2t$$
$$4 - t = -6 + 2t$$
$$-1 + 2t = -1 + t.$$

Since the first equation is solved by $t = 6$, the second by $t = 10/3$, and the third by $t = 0$, no value of t solves all three equations. The two particles never arrive at the same place at the same time, and so they do not collide.

(c) Are there any times t_1 and t_2 such that the position of the first particle at time t_1 is the same as the position of the second particle at time t_2? If so then

$$-1 + t_1 = -7 + 2t_2$$
$$4 - t_1 = -6 + 2t_2$$
$$-1 + 2t_1 = -1 + t_2.$$

We solve the first two equations and get $t_1 = 2$ and $t_2 = 4$. This is a solution for the third equation as well, so the three equations are satisfied by $t_1 = 2$ and $t_2 = 4$. At time $t = 2$ the first particle is at the point $(1, 2, 3)$, and at time $t = 4$ the second is at the same point. The paths cross at the point $(1, 2, 3)$, and the first particle gets there first.

37. For $0 \leq t \leq 2\pi$

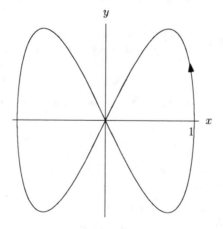

Figure 16.2

41.

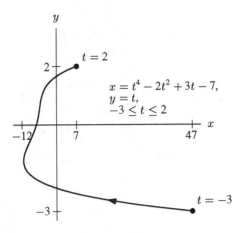

Figure 16.3

The particle starts moving to the left, reverses direction three times, then ends up moving to the right.

Solutions for Section 16.2

1. (a) The first equation gives

$$x = 2 + t, \quad y = 4 + 3t.$$

Eliminating t between this pair of equations gives

$$y - 4 = 3(x - 2),$$
$$y = 3x - 2.$$

The second equation gives

$$x = 1 - 2t, \quad y = 1 - 6t.$$

Eliminating t between this pair of equations gives

$$y - 1 = 3(x - 1),$$
$$y = 3x - 2.$$

Since both parametric equations give rise to the same equation in x and y, they both parameterize the same line.

(b) Slope $= 3$, y-intercept $= -2$.

5. (a) The curve is a spiral as shown in Figure 16.4.

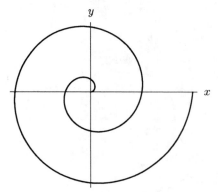

Figure 16.4: The spiral
$x = t\cos t, y = t\sin t$ for $0 \le t \le 4\pi$

(b)   We have:

$$\vec{v}\,(2) \approx \frac{2.001\cos 2.001 - 2\cos 2}{0.001}\vec{i} + \frac{2.001\sin 2.001 - 2\sin 2}{0.001}\vec{j}$$
$$= -2.24\vec{i} + 0.08\vec{j}\,,$$
$$\vec{v}\,(4) \approx \frac{4.001\cos 4.001 - 4\cos 4}{0.001}\vec{i} + \frac{4.001\sin 4.001 - 4\sin 4}{0.001}\vec{j}$$
$$= 2.38\vec{i} - 3.37\vec{j}\,,$$
$$\vec{v}\,(6) \approx \frac{6.001\cos 6.001 - 6\cos 6}{0.001} + \frac{6.001\sin 6.001 - 6\sin 6}{0.001}\vec{j}$$
$$= 2.63\vec{i} + 5.48\vec{j}\,.$$

(c)   Evaluating the exact formula $\vec{v}\,(t) = (\cos t - t\sin t)\vec{i} + (\sin t + t\cos t)\vec{j}$ gives :

$$\vec{v}\,(2) = -2.235\vec{i} + 0.077\vec{j}\,,$$
$$\vec{v}\,(4) = 2.374\vec{i} - 3.371\vec{j}\,,$$
$$\vec{v}\,(6) = 2.637\vec{i} + 5.482\vec{j}\,.$$

See Figure 16.5.

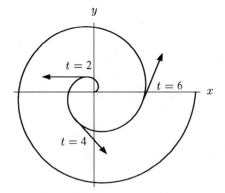

Figure 16.5: The spiral
$x = t\cos t, y = t\sin t$ and three velocity
vectors

9. The velocity vector $\vec{v}$ is given by:

$$\vec{v} = \frac{d}{dt}(t^2 - 2t)\vec{i} + \frac{d}{dt}(t^3 - 3t)\vec{j} + \frac{d}{dt}(3t^4 - 4t^3)\vec{k}$$
$$= (2t - 2)\vec{i} + (3t^2 - 3)\vec{j} + (12t^3 - 12t^2)\vec{k}.$$

The speed is given by:

$$\vec{v} = \sqrt{(2t - 2)^2 + (3t^2 - 3)^2 + (12t^3 - 12t^2)^2}.$$

The particle stops when $2t - 2 = 0$ and $3t^2 - 3 = 0$ and $12t^3 - 12t^2 = 0$. Since these are all satisfied only by $t = 1$, this is the only time that the particle stops.

13. The velocity vector $\vec{v}$ is given by:

$$\vec{v} = \frac{d}{dt}(2 + 3t)\vec{i} + \frac{d}{dt}(4 + t)\vec{j} + \frac{d}{dt}(1 - t)\vec{k} = 3\vec{i} + \vec{j} - \vec{k}.$$

The acceleration vector $\vec{a}$ is given by:

$$\vec{a} = \frac{d\vec{v}}{dt} = \frac{d(3)}{dt}\vec{i} + \frac{d(1)}{dt}\vec{j} - \frac{d(1)}{dt}\vec{k} = \vec{0}$$

17. We have

$$D = \int_0^{2\pi} \sqrt{(-3\sin 3t)^2 + (5\cos 5t)^2} \, dt.$$

We cannot find this integral symbolically, but numerical methods show $D \approx 24.6$.

21. Plotting the positions on the xy plane and noting their times gives the following:

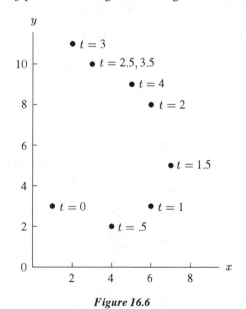

Figure 16.6

(a) We approximate dx/dt by $\Delta x/\Delta t$ calculated between $t = 1.5$ and $t = 2.5$:

$$\frac{dx}{dt} \approx \frac{\Delta x}{\Delta t} = \frac{3 - 7}{2.5 - 1.5} = \frac{-4}{1} = -4.$$

Similarly,

$$\frac{dy}{dt} \approx \frac{\Delta y}{\Delta t} = \frac{10 - 5}{2.5 - 1.5} = \frac{5}{1} = 5.$$

So,

$$\vec{v}(2) \approx -4\vec{i} + 5\vec{j} \quad \text{and} \quad \text{Speed} = \|\vec{v}\| = \sqrt{41}.$$

(b)   The particle is moving vertically at about time $t = 1.5$.
(c)   The particle stops at about time $t = 3$ and reverses course.

25.   Since the acceleration due to gravity is -9.8 m/sec^2, we have $\vec{r}''(t) = -9.8\vec{k}$. Integrating gives

$$\vec{r}'(t) = C_1\vec{i} + C_2\vec{j} + (-9.8t + C_3)\vec{k},$$
$$\vec{r}(t) = (C_1 t + C_4)\vec{i} + (C_2 t + C_5)\vec{j} + (-4.9t^2 + C_3 t + C_6)\vec{k}.$$

The initial condition, $\vec{r}(0) = \vec{0}$, implies that $C_4 = C_5 = C_6 = 0$, thus

$$\vec{r}(t) = C_1 t\vec{i} + C_2 t\vec{j} + (-4.9t^2 + C_3 t)\vec{k}.$$

To find the position vector, we need to find the values of C_1, C_2, and C_3. This we do using the coordinates of the highest point. When the rocket reaches its peak, the vertical component of the velocity is zero, so $-9.8t + C_3 = 0$. Thus, at the highest point, $t = C_3/9.8$. At that time

$$\vec{r}(t) = 1000\vec{i} + 3000\vec{j} + 10000\vec{k},$$

so, for the same value of t:

$$C_1 t = 1000,$$

$$C_2 t = 3000,$$

$$-4.9t^2 + C_3 t = 10,000,$$

Substituting $t = C_3/9.8$ into the third equation gives

$$-4.9\left(\frac{C_3}{9.8}\right)^2 + \frac{C_3^2}{9.8} = 10,000$$
$$C_3^2 = 2(9.8)10,000$$
$$C_3 = 442.7$$

Then $C_1 = \frac{1000}{C_3/9.8} = 22.1$ and $C_2 = \frac{3000}{C_3/9.8} = 66.4$. Thus,

$$\vec{r}(t) = 22.1t\vec{i} + 66.4t\vec{j} + (442.7t - 4.9t^2)\vec{k}.$$

29.   (a)   Consider a time Δt in which a fixed angle at the center is swept out. To find the largest and smallest speeds look for points at which the length of coastline cut off by this beam is greatest, which appears to be C, and smallest, which appears to be E.
      (b)   Now we draw beams of light that sweep out the same area, and again look to see which cuts off the longest length on the shoreline, E, and the smallest, C.
      (c)   Yes, when the light beam is tangential to the shoreline (when the light house is close to A or C).

(d) Consider the right-hand side of the rectangular lake shown in Figure 16.7. We know that $\frac{d}{dt}$(Area of triangle) = constant, where the Area of the triangle = $xk/2$ and the constant k is the distance from L to the right-hand side.

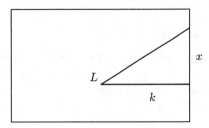

Figure 16.7

Thus,

$$\frac{d}{dt}\left(\frac{xk}{2}\right) = \frac{k}{2}\frac{dx}{dt} = \text{constant},$$

so dx/dt = constant. Therefore the speed is constant along the right-hand side and similarly for the other three sides; it may be different on each side because k varies from side to side. The speed is not defined at the corners.

33. (a) If $\Delta t = t_{i+1} - t_i$ is small enough so that C_i is approximately a straight line, then we can make the linear approximations

$$x(t_{i+1}) \approx x(t_i) + x'(t_i)\Delta t,$$
$$y(t_{i+1}) \approx y(t_i) + y'(t_i)\Delta t,$$
$$z(t_{i+1}) \approx z(t_i) + z'(t_i)\Delta t,$$

and so

$$\begin{aligned}
\text{Length of } C_i &\approx \sqrt{(x(t_{i+1}) - x(t_i))^2 + (y(t_{i+1}) - y(t_i))^2 + (z(t_{i+1}) - z(t_i))^2} \\
&\approx \sqrt{x'(t_i)^2(\Delta t)^2 + y'(t_i)^2(\Delta t)^2 + z'(t_i)^2(\Delta t)^2} \\
&= \sqrt{x'(t_i)^2 + y'(t_i)^2 + z'(t_i)^2}\,\Delta t.
\end{aligned}$$

(b) From point (a) we obtain the approximation

$$\begin{aligned}
\text{Length of } C &= \sum \text{length of } C_i \\
&\approx \sum \sqrt{x'(t_i)^2 + y'(t_i)^2 + z'(t_i)^2}\,\Delta t.
\end{aligned}$$

The approximation gets better and better as Δt approaches zero, and in the limit the sum becomes a definite integral:

$$\begin{aligned}
\text{Length of } C &= \lim_{\Delta t \to 0} \sum \sqrt{x'(t_i)^2 + y'(t_i)^2 + z'(t_i)^2}\,\Delta t \\
&= \int_a^b \sqrt{x'(t)^2 + y'(t)^2 + z'(t)^2}\,dt.
\end{aligned}$$

Solutions for Section 16.3

1. A horizontal disk of radius 5 in the plane $z = 7$.

5. Since $z = r = \sqrt{x^2 + y^2}$, we have a cone around the z-axis. Since $0 \leq r \leq 5$, we have $0 \leq z \leq 5$, so the cone has height and maximum radius of 5.

9. The cross sections of the cylinder perpendicular to the z-axis are circles which are vertical translates of the circle $x^2 + y^2 = a^2$, which is given parametrically by $x = a \cos \theta$, $y = a \sin \theta$. The vector $a \cos \theta \vec{i} + a \sin \theta \vec{j}$ traces out the circle, at any height. We get to a point on the surface by adding that vector to the vector $z \vec{k}$. Hence, the parameters are θ, with $0 \leq \theta \leq 2\pi$, and z, with $0 \leq z \leq h$. The parametric equations for the cylinder are

$$x \vec{i} + y \vec{j} + z \vec{k} = a \cos \theta \vec{i} + a \sin \theta \vec{j} + z \vec{k},$$

which can be written as

$$x = a \cos \theta, \quad y = a \sin \theta, \quad z = z.$$

13. If the planes are parallel, then their normal vectors will also be parallel. The equation of the first plane can be written

$$\vec{r} = 2\vec{i} + 4\vec{j} + \vec{k} + s(\vec{i} + \vec{j} + 2\vec{k}) + t(\vec{i} - \vec{j}).$$

A normal vector to the first plane is $\vec{n}_1 = (\vec{i} + \vec{j} + 2\vec{k}) \times (\vec{i} - \vec{j}) = 2\vec{i} + 2\vec{j} - 2\vec{k}$. The second plane can be written

$$\vec{r} = 2\vec{i} + s(\vec{i} + \vec{k}) + t(2\vec{i} + \vec{j} - \vec{k}).$$

A normal vector to the second plane is $\vec{n}_2 = (\vec{i} + \vec{k}) \times (2\vec{i} + \vec{j} - \vec{k}) = -\vec{i} + 3\vec{j} + \vec{k}$. Since $\vec{n}_1$ and $\vec{n}_2$ are not parallel, neither are the two planes.

17. We use spherical coordinates ϕ and θ as the two parameters. Since the radius is 5, we can take

$$x = 5 \sin \phi \cos \theta, \quad y = 5 \sin \phi \sin \theta, \quad z = 5 \cos \phi.$$

21. Let $(\theta, \pi/2)$ be the original coordinates. If $\theta < \pi$, then the new coordinates will be $(\theta + \pi, \pi/4)$. If $\theta \geq \pi$, then the new coordinates will be $(\theta - \pi, \pi/4)$.

25. (a) The cone of height h, maximum radius a, vertex at the origin and opening upward is shown in Figure 16.8.

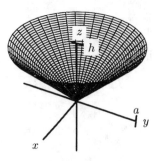

Figure 16.8

By similar triangles, we have

$$\frac{r}{z} = \frac{a}{h},$$

so

$$z = \frac{hr}{a}.$$

Therefore, one parameterization is

$$x = r \cos \theta, \qquad 0 \le r \le a,$$
$$y = r \sin \theta, \qquad 0 \le \theta < 2\pi,$$
$$z = \frac{hr}{a}.$$

(b) Since $r = az/h$, we can write the parameterization in part (a) as

$$x = \frac{az}{h} \cos \theta, \qquad 0 \le z \le h,$$
$$y = \frac{az}{h} \sin \theta, \qquad 0 \le \theta < 2\pi,$$
$$z = z.$$

29. (a) From the first two equations we get:

$$s = \frac{x + y}{2}, \qquad t = \frac{x - y}{2}.$$

Hence the equation of our surface is:

$$z = \left(\frac{x + y}{2} \right)^2 + \left(\frac{x - y}{2} \right)^2 = \frac{x^2}{2} + \frac{y^2}{2},$$

which is the equation of a paraboloid.

The conditions: $0 \le s \le 1, 0 \le t \le 1$ are equivalent to: $0 \le x + y \le 2, 0 \le x - y \le 2$. So our surface is defined by:

$$z = \frac{x^2}{2} + \frac{y^2}{2}, \qquad 0 \le x + y \le 2 \quad 0 \le x - y \le 2$$

(b) The surface is shown in Figure 16.9.

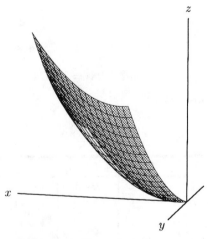

Figure 16.9: The surface $x = s + t$, $y = s - t$, $z = s^2 + t^2$ for $0 \le s \le 1$, $0 \le t \le 1$

33. The plane in which the circle lies is parameterized by

$$\vec{r}(p, q) = x_0\vec{i} + y_0\vec{j} + z_0\vec{k} + p\vec{u} + q\vec{v}.$$

Because $\vec{u}$ and $\vec{v}$ are perpendicular unit vectors, the parameters p and q establish a rectangular coordinate system on this plane exactly analogous to the usual xy-coordinate system, with $(p, q) = (0, 0)$ corresponding to the point (x_0, y_0, z_0). Thus the circle we want to describe, which is the circle of radius a centered at $(p, q) = (0, 0)$, can be parameterized by

$$p = a\cos t, \qquad q = a\sin t.$$

Substituting into the equation of the plane gives the desired parameterization of the circle in 3-space,

$$\vec{r}(t) = x_0\vec{i} + y_0\vec{j} + z_0\vec{k} + a\cos t\vec{u} + a\sin t\vec{v},$$

where $0 \le t \le 2\pi$.

Solutions for Section 16.4

1. The circle $(x - 2)^2 + (y - 2)^2 = 1$.

5. Implicit: $x^2 - 2x + y^2 = 0$, $y < 0$. Explicit: $y = -\sqrt{-x^2 + 2x}$, $0 \le x \le 2$. Parametric: The curve is the lower half of a circle centered at $(1, 0)$ with radius 1, so $x = 1 + \cos t$, $y = \sin t$, for $\pi \le t \le 2\pi$.

9. To find $f_1(-0.02, 0.98)$, we must substitute $x = -0.02$ and $y = 0.98$ into the equation

$$z^3 - 7yz + 6e^x = 0,$$

solve for z, and pick the solution near $z = 2$. Substituting gives

$$z^3 - 6.86z + 6e^{-0.02} = 0.$$

Solving for z gives

$$z = 1.96741, \quad z = 1.00551, \quad z = -2.97292.$$

We pick $z = 1.96741$, because f_1 gives values of z near $z = 2$. Thus,

$$f(-0.02, 0.98) = 1.96741.$$

This gives us the first entry in Table 16.1.

TABLE 16.1 *Values of $f_1(x, y)$*

y/x	−0.02	−0.02	0.00	0.01	0.02
0.98	1.96741	1.95477	1.94158	1.92778	1.9133
0.99	1.99580	1.98385	1.97142	1.95847	1.94492
1.00	2.02312	2.01177	2	1.98776	1.97501
1.01	2.04949	2.03868	2.02747	2.01586	2.00379
1.02	2.07502	2.06468	2.05398	2.04292	2.03145

13. (a) The plane $2(x-3)+4(y-5)+5(z-7) = 0$ will be tangent to the graph of the surface $f(x,y,z) = 0$ at the point $(3,5,7)$. Near the point $(3,5,7)$ the graph will be very well approximated by this plane. We do not have any information about the graph far from the point $(3,5,7)$.

  (b) The solutions of the equation $f(x,y,z)$ near $(3,5,7)$ will be well approximated by the solutions of the linear equation $2(x-3) + 4(y-5) + 5(z-7) = 0$. Solving this linear equation for z, we get $z = 7 - (2/5)(x-3) - (4/5)(y-5)$. We conclude that for every (x,y) near $(3,5)$ there is one and only one solution of the equation $f(x,y,z) = 0$ that is near 7. That solution is well approximated by the formula $z \approx 7 - (2/5)(x-3) - (4/5)(y-5)$.

Solutions for Section 16.5

1. The north-south distance between Alexandria and Syene corresponds to $1/50$ of a full circle and measures 5,000 stadia. Thus if C is the circumference of the earth, we have

$$\frac{5000}{C} = \frac{1}{50}$$

giving us

$$C = 250{,}000 \text{ stadia.}$$

Using the fact that 1 stadium $= 185$ meters, we get

$$C = 250{,}000 \text{ stadia} \cdot 185 \text{ meters/stadia} = 46{,}250{,}000 \text{ meters} \approx 46{,}000 \text{ km.}$$

(In fact, the circumference is about 40,000 km.)

5. Differentiating the equations given, we see that

$$\frac{d^2x}{dt^2} = \frac{du}{dt} = \frac{-kx}{x^2+y^2}.$$

Similarly

$$\frac{d^2y}{dt^2} = \frac{dv}{dt} = \frac{-ky}{x^2+y^2}.$$

The acceleration vector $\vec{a}$ is given by

$$\vec{a} = \frac{d^2x}{dt^2}\vec{i} + \frac{d^2y}{dt^2}\vec{j} = \frac{-kx}{x^2+y^2}\vec{i} - \frac{ky}{x^2+y^2}\vec{j} = \frac{-k}{x^2+y^2}(x\vec{i}+y\vec{j}).$$

Thus,

$$\vec{a} = -\frac{k\vec{r}}{r^2}.$$

So $\vec{a}$ points toward the origin and $\|\vec{a}\| = k/r$.

Orbits are not always closed for solutions to differential equations representing a general centripetal acceleration.

Solutions for Chapter 16 Review

1. $x = t, y = 5$.

5. The vector $(\vec{i} + 2\vec{j} + 5\vec{k}) - (2\vec{i} - \vec{j} + 4\vec{k}) = -\vec{i} + 3\vec{j} + \vec{k}$ is parallel to the line, so a possible parameterization is

$$x = 2 - t, \quad y = -1 + 3t, \quad z = 4 + t.$$

9. Since the circle has radius 3, the equation must be of the form $x = 3\cos t, y = 5, z = 3\sin t$. But since the circle is being viewed from farther out on the y-axis, the circle we have now would be seen going clockwise. To correct this, we add a negative to the third component, giving us the equation $x = 3\cos t, y = 5, z = -3\sin t$.

13. (a) $f_x = \dfrac{[2x(x^2 + y^2) - 2x(x^2 - y^2)]}{(x^2 + y^2)^2} = \dfrac{4xy^2}{(x^2 + y^2)^2}.$

$f_y = \dfrac{[-2y(x^2 + y^2) - 2y(x^2 - y^2)]}{(x^2 + y^2)^2} = \dfrac{-4yx^2}{(x^2 + y^2)^2}.$

$\nabla f(1, 1) = \vec{i} - \vec{j}$, i.e., south-east.

(b) We need a vector $\vec{u}$ such that $\nabla f(1, 1) \cdot \vec{u} = 0$, i.e., such that $(\vec{i} - \vec{j}) \cdot \vec{u} = 0$. The vector $\vec{u} = \vec{i} + \vec{j}$ clearly works; so does $\vec{u} = -\vec{i} - \vec{j}$. Dividing by the length to get a unit vector, we have $\vec{u} = \frac{1}{\sqrt{2}}\vec{i} + \frac{1}{\sqrt{2}}\vec{j}$ or $\vec{u} = -\frac{1}{\sqrt{2}}\vec{i} - \frac{1}{\sqrt{2}}\vec{j}$.

(c) f is a function of x and y, which in turn are functions of t. Thus, the chain rule can be used to show how f changed with t.

$$\frac{df}{dt} = \frac{\partial f}{\partial x} \cdot \frac{dx}{dt} + \frac{\partial f}{\partial y} \cdot \frac{dy}{dt} = \frac{4xy^2}{(x^2 + y^2)^2} \cdot 2e^{2t} - \frac{4x^2 y}{(x^2 + y^2)^2} \cdot (6t^2 + 6).$$

At $t = 0, x = 1, y = 1$; so, $\dfrac{df}{dt} = \dfrac{4}{4} \cdot 2 - \dfrac{4}{4} \cdot 6 = -4$.

17. The plot looks like Figure 16.10.

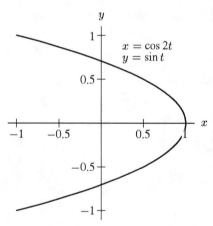

Figure 16.10

which does appear to be part of a parabola. To prove that it is, we note that we have

$$x = \cos 2t$$

$$y = \sin t$$

and must somehow find a relationship between x and y. Recall the trig identity

$$\cos 2t = 1 - 2\sin^2 t.$$

Thus we have $x = 1 - 2y^2$, which is a parabola lying along the x-axis, for $-1 \le y \le 1$.

21. Set up the coordinate system as shown in Figure 16.11.

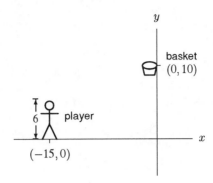

Figure 16.11

(a) We separate the initial velocity vector into its x and y components.

$$V_x = V \cos A$$

$$V_y = V \sin A.$$

Since there is no force acting in the x direction, the x-coordinate of the basketball is just

$$x = (V \cos A)t - 15.$$

For the y-coordinate, we know that
$$y''(t) = -32,$$

so

$$y'(t) = -32t + C_1$$

and

$$y(t) = -16t^2 + C_1 t + C_2.$$

We also know that $y'(0) = V \sin A$ and $y(0) = 6$. Substituting these values in, we get $C_1 = V \sin A$, $C_2 = 6$ and thus
$$y = -16t^2 + (V \sin A)t + 6.$$

(b) Many pairs of V and A work. For example, $V = 26, A = 60°, V = 32, A = 30°$.

(c) Now that we have the equations, we need to find a relationship between V and A that ensures that the basketball goes through the hoop (i.e., the curve passes through $(0, 10)$). So we set

$$x = (V \cos A)t - 15 = 0$$

$$y = -16t^2 + (V \sin A)t + 6 = 10.$$

From the first equation, we get $t = \frac{15}{V \cos A}$. Then we substitute that into the second equation:

$$-16\left(\frac{15}{V \cos A}\right)^2 + (V \sin A)(\frac{15}{V \cos A}) = 4$$

$$-\frac{3600}{V^2 \cos^2 A} + 15 \tan A = 4$$

$$V^2 = \frac{3600}{\cos^2 A(15\tan A - 4)}.$$

Keeping in mind that $\tan^2\theta + 1 = \frac{1}{\cos^2\theta}$, one has:

$$V^2 = \frac{3600(1 + \tan^2 A)}{15\tan A - 4}.$$

We can minimize V by minimizing V^2 (since $V > 0$).

$$\frac{d(V^2)}{dA} = \frac{2\tan A(15\tan A - 4) - 15(\tan^2 A + 1)}{(15\tan A - 4)^2} \cdot \frac{3600}{\cos^2 A} = 0$$

$$\frac{3600}{\cos^2 A}\left[\frac{15\tan^2 A - 8\tan A - 15}{(15\tan A - 4)^2}\right] = 0$$

$$15\tan^2 A - 8\tan A - 15 = 0$$

$$\tan A = \frac{8 + \sqrt{964}}{30}$$

$$\approx 1.30$$

$$A \approx 52°.$$

25. To parameterize a plane we must find one point in the plane and two nonparallel vectors in the plane. Since there is an infinity of possibilities, there are many possible parameterizations.

Points (x, y, z) in the plane can be found by picking values for two coordinates and using the equation of the plane to solve for the third. For example, with $x = 2$ and $y = 3$ we have $3(2) + 4(3) + 5z = 10$, and so $z = -1.6$. Thus the point $P = (2, 3, -1.6)$ is in the plane.

Vectors in the plane can be found as displacement vectors between two points in the plane. To get two displacement vectors, we need three points in the plane. If $x = y = 0$, then $z = 2$, giving the point $Q = (0, 0, 2)$ in the plane. And if $x = z = 0$, the $y = 2.5$, giving the point $R = (0, 2.5, 0)$ in the plane. The vectors $\vec{v}_1 = \overrightarrow{PQ} = -2\vec{i} - 3\vec{j} + 3.6\vec{k}$ and $\vec{v}_2 = \overrightarrow{PR} = -2\vec{i} - 0, 5\vec{j} + 1.6\vec{k}$ lie in the plane. Hence one parameterization of the plane is

$$\vec{r} = 2\vec{i} + 3\vec{j} - 1.6\vec{k} + s\vec{v}_1 + t\vec{v}_2$$
$$= (2 - 2s - 2t)\vec{i} + (3 - 3s - 0.5t)\vec{j} + (-1.6 + 3.6s + 1.6t)\vec{k}$$

or, equivalently

$$x = 2 - 2s - 2t$$
$$y = 3 - 3s - 0.5t$$
$$z = -1.6 + 3.6s + 1.6t.$$

Notice how the parametric equations satisfy the equation of the plane:

$$3(2 - 2s - 2t) + 4(3 - 3s - 0.5t) + 5(-1.6 + 3.6s + 1.6t) = 10.$$

29. We are given the conditions $f(0) = 15$ and $f(20) = 1$ and asked to solve for a and b. The equations are

$$b = 15a^m \quad b = (20 + a)^m$$

from which

$$15^{1/m}a = 20 + a$$

and so

$$a = \frac{20}{(15^{1/m} - 1)}.$$

The values for a and b corresponding to $m = 0.5, 0.7$, and 1 are given in Table 16.2.

TABLE 16.2

m	0.5	0.7	1.0
a	0.0893	0.427	1.43
b	4.48	8.26	21.4

Figure 16.12 shows the graphs of f when $m = 0.5, 0.7$, and 1.

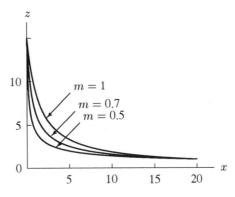

Figure 16.12

Notice that the surface obtained by rotating the graph of $z = f(x)$ in the xz-plane, for $x > 0$, about the x-axis will be more "flared" when $m = 0.5$ that it is for $m = 1$.

33. Suppose we choose a parameterization in which the particle is at the point (a, b, c) when $t = 0$ and at the point (A, B, C) when $t = 1$. Then the x-coordinate changes by $(A - a)$ in unit time, the y-coordinate changes by $(B - b)$, and the z-coordinate by $(C - c)$. Thus, parametric equations for the line of sight are

$$x = a + t(A - a), \quad y = b + t(B - b), \quad z = c + t(C - c).$$

We now substitute these expressions into the equation $Ax + By + Cz = D$ and solve for t:

$$A(a + t(A - a)) + B(b + t(B - b)) + C(c + t(C - c)) = D$$
$$Aa + Bb + Cc + t(A^2 + B^2 + C^2 - Aa - Bb - Cc) = D.$$

Writing $E = Aa + Bb + Cc$ and $F = A^2 + B^2 + C^2$, we get

$$t = \frac{D - E}{F - E}.$$

Then we have:

$$x = a + \frac{D-E}{F-E}(A-a)$$
$$y = b + \frac{D-E}{F-E}(B-b)$$
$$z = c + \frac{D-E}{F-E}(C-c)$$

The problem of converting the xyz-coordinates of the point of intersection into two-dimensional screen coordinates is considered in Problem 34.

CHAPTER SEVENTEEN

1. Notice that for a repulsive force, the vectors point outward, away from the particle at the origin, for an attractive force, the vectors point toward the particle. So we can match up the vector field with the description as follows:

   (a) IV
   (b) III
   (c) I
   (d) II

5.

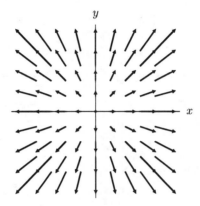

Figure 17.1: $\vec{F}(\vec{r}) = 2\vec{r}$

9. $\vec{V} = -y\vec{i}$

13. $\vec{V} = -y\vec{i} + x\vec{j}$

17.

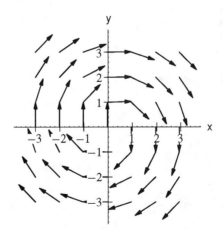

The position vector at each point is $\vec{r} = x\vec{i} + y\vec{j}$. We want to find $\vec{F}(x, y) = A\vec{i} + B\vec{j}$ such that $\vec{F} \cdot \vec{r} = Ax + By = 0$. One possible answer is let $A = y$ and $B = -x$. So $\vec{F}(x, y) = y\vec{i} - x\vec{j}$. Since the vectors are of unit length, we get $\vec{F}(x, y) = \dfrac{y\vec{i} - x\vec{j}}{\sqrt{x^2 + y^2}}$.

Solutions for Section 17.2

1. Since $x'(t) = 3$ and $y'(t) = 0$, we have $x = 3t + x_0$ and $y = y_0$. Thus, the solution curves are $y = $ constant.

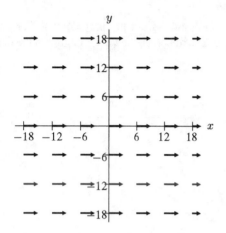

Figure 17.2: The field $\vec{v} = 3\vec{i}$

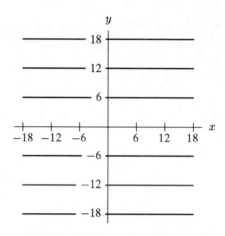

Figure 17.3: The flow $y = $ constant

5.

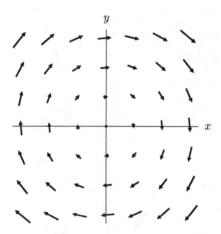

Figure 17.4: $\vec{v}(t) = y\vec{i} - x\vec{j}$

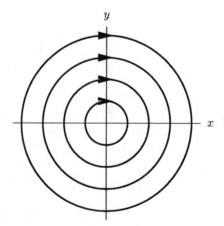

Figure 17.5: The flow $x = a \sin t$,
$y = a \cos t$

As

$$\vec{v}(t) = \frac{dx}{dt}\vec{i} + \frac{dy}{dt}\vec{j},$$

the system of differential equations is

$$\begin{cases} \frac{dx}{dt} = y \\ \frac{dy}{dt} = -x. \end{cases}$$

Since

$$\frac{dx(t)}{dt} = \frac{d}{dt}[a \sin t] = a \cos t = y(t)$$

and

$$\frac{dy(t)}{dt} = \frac{d}{dt}[a \cos t] = -a \sin t = -x(t),$$

the given flow satisfies the system. By eliminating the parameter t in $x(t)$ and $y(t)$, the solution curves obtained are $x^2 + y^2 = a^2$.

9. The directions of the flow lines are as shown.

   (a) III
   (b) I
   (c) II
   (d) V
   (e) VI
   (f) IV

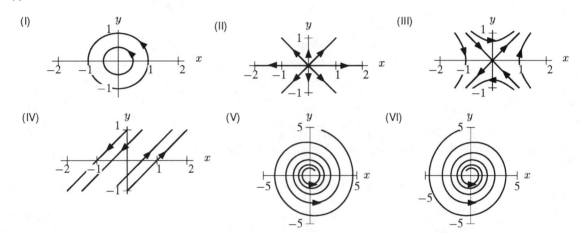

Solutions for Chapter 17 Review

1. (a) A vector field associates a vector to every point in a region of the space. In other words, a vector field is a vector-valued function of position given by $\vec{v} = \vec{f}(\vec{r}) = \vec{f}(x, y, z)$

   (b) (i) Yes, $\vec{r} + \vec{a} = (x + a_1)\vec{i} + (y + a_2)\vec{j} + (z + a_3)\vec{k}$ is a vector-valued function of position.

   (ii) No, $\vec{r} \cdot \vec{a}$ is a scalar.

   (iii) Yes.

   (iv) $x^2 + y^2 + z^2$ is a scalar.

5. (a) Since $\vec{F} = \frac{\vec{r}}{\|\vec{r}\|^3}$, the magnitude of $\vec{F}$ is given by

$$\|\vec{F}\| = \frac{\|\vec{r}\|}{\|\vec{r}\|^3} = \frac{1}{\|\vec{r}\|^2}.$$

Now $\vec{r} = x\vec{i} + y\vec{j} + z\vec{k}$, so the magnitude of $\vec{r}$ is given by

$$\|\vec{r}\| = \sqrt{x^2 + y^2 + z^2}.$$

Thus,

$$\|\vec{F}\| = \frac{1}{\|\vec{r}\|^2} = \frac{1}{x^2 + y^2 + z^2}.$$

(b) $\vec{F} \cdot \vec{r} = \frac{\vec{r}}{\|\vec{r}\|^3} \cdot \vec{r} = \frac{\|\vec{r}\|^2}{\|\vec{r}\|^3} = \frac{1}{\|\vec{r}\|} = \frac{1}{\sqrt{x^2+y^2+z^2}}.$

(c) A unit vector parallel to $\vec{F}$ and pointing in the same direction is given by $\vec{U} = \frac{\vec{F}}{\|\vec{F}\|}.$

$\vec{F} = \frac{\vec{r}}{\|\vec{r}\|^3}$, and $\|\vec{F}\| = \frac{1}{\|\vec{r}\|^2}$. Putting these into the expression for $\vec{U}$ we have

$$\vec{U} = \frac{\vec{F}}{\|\vec{F}\|} = \frac{\frac{\vec{r}}{\|\vec{r}\|^3}}{\frac{1}{\|\vec{r}\|^2}} = \frac{\vec{r}}{\|\vec{r}\|}$$

$$= \frac{x}{\sqrt{x^2+y^2+z^2}}\vec{i} + \frac{y}{\sqrt{x^2+y^2+z^2}}\vec{j} + \frac{z}{\sqrt{x^2+y^2+z^2}}\vec{k}.$$

(d) A unit vector parallel to $\vec{F}$ and pointing in the opposite direction is given by:

$$\vec{V} = -\frac{\vec{F}}{\|\vec{F}\|} = -\frac{\vec{r}}{\|\vec{r}\|}$$

$$= \frac{-x}{\sqrt{x^2+y^2+z^2}}\vec{i} + \frac{-y}{\sqrt{x^2+y^2+z^2}}\vec{j} + \frac{-z}{\sqrt{x^2+y^2+z^2}}\vec{k}.$$

(e) If $\vec{r} = \cos t\vec{i} + \sin t\vec{j} + \vec{k}$, then $\|\vec{r}\| = \sqrt{\cos^2 t + \sin^2 t + 1} = \sqrt{2}$.
So, $\vec{F} = \frac{\vec{r}}{\|\vec{r}\|^3} = \frac{\cos t}{\sqrt{8}}\vec{i} + \frac{\sin t}{\sqrt{8}}\vec{j} + \frac{1}{\sqrt{8}}\vec{k} = \frac{\cos t}{2\sqrt{2}}\vec{i} + \frac{\sin t}{2\sqrt{2}}\vec{j} + \frac{1}{2\sqrt{2}}\vec{k}.$

(f) We know that $\vec{F} \cdot \vec{r} = \frac{1}{\|\vec{r}\|}$, so if $\vec{r} = \cos t\vec{i} + \sin t\vec{j} + \vec{k}$, $\vec{F} \cdot \vec{r} = \frac{1}{\sqrt{2}}.$

9. This corresponds to area D in Figure 17.6.

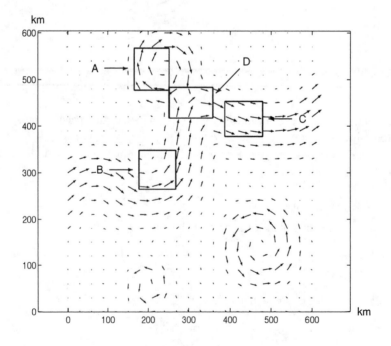

Figure 17.6

CHAPTER EIGHTEEN

Solutions for Section 18.1

1. Positive, because the vectors are longer on the portion of the path that goes in the same direction as the vector field.

5. Since it appears that C_1 is everywhere perpendicular to the vector field, all of the dot products in the line integral are zero, hence $\int_{C_1} \vec{F} \cdot d\vec{r} \approx 0$. Along the path C_2 the dot products of $\vec{F}$ with $\Delta\vec{r}_i$ are all positive, so their sum is positive and we have $\int_{C_1} \vec{F} \cdot d\vec{r} < \int_{C_2} \vec{F} \cdot d\vec{r}$. For C_3 the vectors $\Delta\vec{r}_i$ are in the opposite direction to the vectors of $\vec{F}$, so the dot products $\vec{F} \cdot \Delta\vec{r}_i$ are all negative; so, $\int_{C_3} \vec{F} \cdot d\vec{r} < 0$. Thus, we have

$$\int_{C_3} \vec{F} \cdot d\vec{r} < \int_{C_1} \vec{F} \cdot d\vec{r} < \int_{C_2} \vec{F} \cdot d\vec{r}$$

9. Since it does not depend on y, this vector field is constant along vertical lines, $x = $ constant. Now let us consider two points P and Q on C_1 which lie on the same vertical line. Because C_1 is symmetric with respect to the x-axis, the tangent vectors at P and Q will be symmetric with respect to the vertical axis so their sum is a vertical vector. But $\vec{F}$ has only horizontal component and thus $\vec{F} \cdot (\Delta\vec{r}(P) + \Delta\vec{r}(Q)) = 0$. As $\vec{F}$ is constant along vertical lines (so $\vec{F}(P) = \vec{F}(Q)$), we obtain

$$\vec{F}(P) \cdot \Delta\vec{r}(P) + \vec{F}(Q) \cdot \Delta\vec{r}(Q) = 0.$$

Summing these products and making $\|\Delta\vec{r}\| \to 0$ gives us

$$\int_{C_1} \vec{F} \cdot d\vec{r} = 0.$$

The same thing happens on C_3, so $\int_{C_3} \vec{F} \cdot d\vec{r} = 0$.

Now let P be on C_2 and Q on C_4 lying on the same vertical line. The respective tangent vectors are symmetric with respect to the vertical axis hence they add up to a vertical vector and a similar argument as before gives

$$\vec{F}(P) \cdot \Delta\vec{r}(P) + \vec{F}(Q) \cdot \Delta\vec{r}(Q) = 0$$

and

$$\int_{C_2} \vec{F} \cdot d\vec{r} + \int_{C_4} \vec{F} \cdot d\vec{r} = 0$$

and so

$$\int_{C} \vec{F} \cdot d\vec{r} = 0.$$

See Figure 18.1.

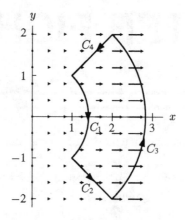

Figure 18.1

13.   At every point, the vector field is parallel to segments $\Delta \vec{r} = \Delta x \vec{i}$ of the curve. Thus,

$$\int_C \vec{F} \cdot d\vec{r} = \int_2^6 x\vec{i} \cdot dx\vec{i} = \int_2^6 x \, dx = \left. \frac{x^2}{2} \right|_2^6 = 16.$$

17.   See Figure 18.2. The example chosen is the vector field $\vec{F}(x, y) = y\vec{j}$ and the path C is the line from $(0, -1)$ to $(0, 1)$. Since the vectors are symmetric about the x-axis, the dot products $\vec{F} \cdot \Delta \vec{r}$ cancel out along C to give 0 for the line integral. Many other answers are possible.

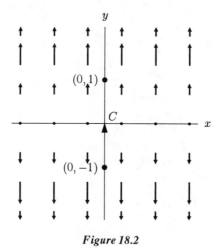

Figure 18.2

21.   (a)   Different closed curves C, will give different values for the line integral

$$\int_C \vec{F} \cdot d\vec{r}.$$

(b)   The value of the line integral takes on different values, depending on the path taken.

25. Pick any closed curve C. Choose two distinct points P_1, P_2 on C. Let C_1, C_2 be the two curves from P_1 to P_2 along C. See Figure 18.3. Let $-C_2$ be the same as C_2, except in the opposite direction. Thus, $C_1 - C_2 = C$. Therefore,

$$\int_C \vec{F} \cdot d\vec{r} = \int_{C_1 - C_2} \vec{F} \cdot d\vec{r} = \int_{C_1} \vec{F} \cdot d\vec{r} - \int_{C_2} \vec{F} \cdot d\vec{r}$$

since C_2 and $-C_2$ differ only in direction. But C_1 and C_2 have the same endpoints (P_1 and P_2) and same direction (P_1 to P_2), so by assumption we have $\int_{C_1} \vec{F} \cdot d\vec{r} = \int_{C_2} \vec{F} \cdot d\vec{r}$. Therefore,

$$\int_C \vec{F} \cdot d\vec{r} = \int_{C_1} \vec{F} \cdot d\vec{r} - \int_{C_2} \vec{F} \cdot d\vec{r} = 0.$$

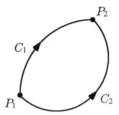

Figure 18.3

Solutions for Section 18.2

1. The parameterization is given, so

$$\int_C \vec{F} \cdot d\vec{r} = \int_2^4 \vec{F}(2t, t^3) \cdot (2\vec{i} + 3t^2\vec{j})\, dt$$

$$= \int_2^4 [(\ln(t^3)\vec{i} + \ln(2t)\vec{j}] \cdot (2\vec{i} + 3t^2\vec{j})\, dt$$

$$= \int_2^4 (2\ln(t^3) + 3t^2\ln(2t))\, dt$$

$$= \int_2^4 (6\ln(t) + 3t^2\ln(2t))\, dt \qquad \text{since } \ln(t^3) = 3\ln(t).$$

This integral can be computed numerically, or using integration by parts or the integral table, giving

$$\int_C \vec{F} \cdot d\vec{r} = \int_2^4 (6\ln(t) + 3t^2\ln(2t))\, dt$$

$$= (6(t\ln(t) - t) + t^3\ln(2t) - t^3/3)\Big|_2^4$$

$$= 240\ln 2 - \frac{136}{3} - (28\ln 2 - \frac{44}{3})$$

$$= 212\ln 2 - 92/3 \approx 116.28.$$

The expression containing $\ln 2$ was obtained using the properties of the natural log.

5. The portion of the ellipse can be parameterized by $(2\cos t, \sin t)$, for $0 \leq t \leq \pi/2$, but this gives a *counterclockwise* orientation. Thus $t = \pi/2$ gives the beginning of the curve and $t = 0$ gives the end, so

$$\int_C \vec{F} \cdot d\vec{r} = \int_{\pi/2}^{0} \vec{F}(2\cos t, \sin t) \cdot (-2\sin t\vec{i} + \cos t\vec{j})\, dt$$

$$= -\int_0^{\pi/2} (e^{2\cos t}\vec{i} + e^{\sin t}\vec{j}) \cdot (-2\sin t\vec{i} + \cos t\vec{j})\, dt$$

$$= -\int_0^{\pi/2} (-2e^{2\cos t}\sin t + e^{\sin t}\cos t)\, dt$$

$$= -(e^{2\cos t} + e^{\sin t})\Big|_0^{\pi/2}$$

$$= -[e^0 + e^1 - (e^2 + e^0)] = e^2 - e.$$

9. Since C is given by $\vec{r} = \cos t\vec{i} + \sin t\vec{j} + t\vec{k}$, we have $\vec{r}'(t) = -\sin t\vec{i} + \cos t\vec{j} + \vec{k}$. Thus,

$$\int_C \vec{F} \cdot d\vec{r} = \int_0^{4\pi} (-\sin t\vec{i} + \cos t\vec{j} + 5\vec{k}) \cdot (-\sin t\vec{i} + \cos t\vec{j} + \vec{k})\, dt$$

$$= \int_0^{4\pi} (\sin^2 t + \cos^2 t + 5)\, dt = \int_0^{4\pi} 6\, dt = 24\pi.$$

13. (a) Since $\vec{r}(t) = t\vec{i} + t^2\vec{j}$, we have $\vec{r}'(t) = \vec{i} + 2t\vec{j}$. Thus,

$$\int_C \vec{F} \cdot d\vec{r} = \int_0^1 \vec{F}(t, t^2) \cdot (\vec{i} + 2t\vec{j})\, dt$$

$$= \int_0^1 [(3t - t^2)\vec{i} + t\vec{j}] \cdot (\vec{i} + 2t\vec{j})\, dt$$

$$= \int_0^1 (3t + t^2)\, dt$$

$$= (\frac{3t^2}{2} + \frac{t^3}{3})\Big|_0^1$$

$$= \frac{3}{2} + \frac{1}{3} - (0 + 0) = \frac{11}{6}$$

(b) Since $\vec{r}(t) = t^2\vec{i} + t\vec{j}$, we have $\vec{r}'(t) = 2t\vec{i} + \vec{j}$. Thus,

$$\int_C \vec{F} \cdot d\vec{r} = \int_0^1 \vec{F}(t^2, t) \cdot (2t\vec{i} + \vec{j})\, dt$$

$$= \int_0^1 [(3t^2 - t)\vec{i} + t^2\vec{j}] \cdot (2t\vec{i} + \vec{j})\, dt$$

$$= \int_0^1 (6t^3 - t^2)\, dt$$

$$= (\frac{3t^4}{2} - \frac{t^3}{3})\Big|_0^1$$

$$= \frac{3}{2} - \frac{1}{3} - (0 - 0) = \frac{7}{6}$$

17. The integral corresponding to $A(t) = (t, t)$ is

$$\int_0^1 3t \, dt.$$

The integral corresponding to $B(t) = (2t, 2t)$ is

$$\int_0^{1/2} 12t \, dt.$$

The substitution $s = 2t$ has $ds = 2\, dt$ and $s = 0$ when $t = 0$ and $s = 1$ when $t = 1/2$. Thus, substituting $t = \dfrac{s}{2}$ into the integral corresponding to $B(t)$ gives

$$\int_0^{1/2} 12t \, dt = \int_0^1 12(\tfrac{s}{2})(\tfrac{1}{2}\, ds) = \int_0^1 3s \, ds.$$

The integral on the right-hand side is now the same as the integral corresponding to $A(t)$. Therefore we have

$$\int_0^{1/2} 12t \, dt = \int_0^1 3s \, ds = \int_0^1 3t \, dt.$$

Alternatively, a similar calculation shows that the substitution $t = 2w$ converts the integral corresponding to $A(t)$ into the integral corresponding to $B(t)$.

Solutions for Section 18.3

1. The vector field $\vec{F}$ points radially outward, and so is everywhere perpendicular to A; thus, $\int_A \vec{F} \cdot d\vec{r} = 0$.

 Along the first half of B, the terms $\vec{F} \cdot \Delta \vec{r}$ are negative; along the second half the terms $\vec{F} \cdot \Delta \vec{r}$ are positive. By symmetry the positive and negative contributions cancel out, giving a Riemann sum and a line integral of 0.

 The line integral is also 0 along C, by cancellation. Here the values of $\vec{F}$ along the x-axis have the same magnitude as those along the y-axis. On the first half of C the path is traversed in the opposite direction to $\vec{F}$; on the second half of C the path is traversed in the same direction as $\vec{F}$. So the two halves cancel.

5. No. Suppose there were a function f such that grad $f = \vec{F}$. Then we would have

$$\frac{\partial f}{\partial x} = \frac{-z}{\sqrt{x^2 + z^2}}.$$

Hence we would have

$$\frac{\partial^2 f}{\partial y \partial x} = \frac{\partial}{\partial y}(\frac{-z}{\sqrt{x^2 + z^2}}) = 0.$$

In addition, since grad $f = \vec{F}$, we have that

$$\frac{\partial f}{\partial y} = \frac{y}{\sqrt{x^2 + z^2}}.$$

Thus we also know that

$$\frac{\partial^2 f}{\partial x \partial y} = \frac{\partial}{\partial x}\left(\frac{y}{\sqrt{x^2 + y^2}}\right) = -xy(x^2 + z^2)^{-3/2}.$$

Notice that

$$\frac{\partial^2 f}{\partial y \partial x} \neq \frac{\partial^2 f}{\partial x \partial y}.$$

Since we expect $\frac{\partial^2 f}{\partial y \partial x} = \frac{\partial^2 f}{\partial x \partial y}$, we have got a contradiction. The only way out of this contradiction is to conclude there is no function f with grad $f = \vec{F}$. Thus $\vec{F}$ is not a gradient vector field.

9. Since $\vec{F} = \text{grad}(y \ln(x+1))$, we evaluate the line integral using the Fundamental Theorem of Line Integrals:

$$\int_C \vec{F} \cdot d\vec{r} = y \ln(x+1) \Big|_{(0,0)}^{(3/\sqrt{2}, 3/\sqrt{2})} = \frac{3}{\sqrt{2}} \ln\left(\frac{3}{\sqrt{2}} + 1\right) - 0 \ln 1 = \frac{3}{\sqrt{2}} \ln\left(\frac{3}{\sqrt{2}} + 1\right).$$

13. (a) Three possible paths are shown in Figure 18.4.

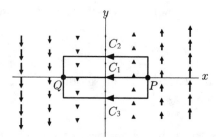

Figure 18.4

Since $\vec{F}$ is perpendicular to the horizontal axis everywhere, $\vec{F} \cdot d\vec{r} = 0$ along C_1.

Since C_2 starts out in the direction of $\vec{F}$, the first leg of C_2 will have a positive line integral. The second horizontal part of C_2 will have a 0 line integral, and the third leg that ends at Q will have a positive line integral. Thus the line integral along C_2 is positive.

A similar argument shows that the line integral along $C_3 < 0$.

(b) No, $\vec{F}$ is not a gradient field, since the line integrals along these three paths joining P and Q do not have the same value.

17. You can easily come up with counterexamples: suppose that $\vec{F} \neq \vec{G}$ but that both are gradient fields. For example, $\vec{F} = \vec{i}$ and $\vec{G} = \vec{j}$. Then, if C is a closed curve, the line integral around C of both $\vec{F}$ and $\vec{G}$ will equal to zero. But this does not mean that $\vec{F} = \vec{G}$.

21. (a) The level surfaces are horizontal planes given by $gz = c$, so $z = c/g$. The potential energy increases with the height above the earth. This means that more energy is stored as "potential to fall" as height increases.

(b) The gradient of ϕ points upward (in the direction of increasing potential energy), so $\nabla \phi = g\vec{k}$. The gravitational force acts toward the earth in the direction of $-\vec{k}$. So, $\vec{F} = -g\vec{k}$. The negative sign represents the fact that the gravitational force acts in the direction of the decreasing potential energy.

Solutions for Section 18.4

1. The drawing of the contour diagrams fitting this gradient field would look like Figure 18.5:

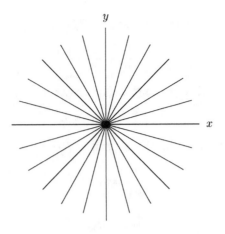

Figure 18.5

This diagram could not be the contour diagram because the origin is on all contours. This means that $f(0, 0)$ would have to take on more than one value, which is impossible. At a point P other than the origin, we have the same problem. The values on the contours increase as you go counterclockwise around, since the gradient vector points in the direction of greatest increase of a function. But, starting at P, and going all the way around the origin, you would eventually get back to P again, and with a larger value of f, which is impossible.

5. The domain of the vector field $\vec{F}(x, y) = y\vec{i} + y\vec{j}$ is the whole xy-plane. In order to see if $\vec{F}$ is a gradient let us apply the curl test:

$$\frac{\partial F_1}{\partial y} = 1$$

and

$$\frac{\partial F_2}{\partial x} = 0$$

So $\vec{F}$ is not the gradient of any function.

9. The domain of the vector field $\vec{F} = \dfrac{\vec{i}}{x} + \dfrac{\vec{j}}{y} + \dfrac{\vec{k}}{z}$ is the set of points (x, y, z) in the three space such that $x \neq 0$, $y \neq 0$ and $z \neq 0$. This is what is left in the three space after removing the coordinate planes.

This domain has the property that every closed curve is the boundary of a surface entirely contained in it, hence we can apply the curl test.

$$\text{curl } \vec{F} = \begin{vmatrix} \vec{i} & \vec{j} & \vec{k} \\ \frac{\partial}{\partial x} & \frac{\partial}{\partial y} & \frac{\partial}{\partial z} \\ \frac{1}{x} & \frac{1}{y} & \frac{1}{z} \end{vmatrix}$$

So curl $\vec{F} = \vec{0}$ and thus $\vec{F}$ is the gradient of a function f. In order to compute f we first integrate

$$\frac{\partial f}{\partial x} = \frac{1}{x}$$

with respect to x, thinking of y and z as constants. We get

$$f(x, y, z) = \ln|x| + C(y, z)$$

Differentiating with respect to y and using the fact that $\partial f / \partial y = 1/y$ gives

$$\frac{\partial f}{\partial y} = \frac{\partial C}{\partial y} = \frac{1}{y}$$

We integrate this relation with respect to y thinking of z as a constant. We get

$$f(x, y, z) = \ln |xy| + K(z)$$

Differentiating with respect to z and using the fact that $\partial f / \partial z = 1/z$ gives

$$\frac{\partial f}{\partial z} = K'(z) = \frac{1}{z}$$

Now we integrate with respect to z and get

$$f(x, y, z) = \ln A|xyz|$$

where A is a positive constant.

13. Using $\vec{F} = x\vec{j} = a\cos t\vec{j}$ and $\vec{r}'(t) = -a\sin t\vec{i} + b\cos t\vec{j}$, we have

$$\begin{aligned} A &= \int_C \vec{F} \cdot d\vec{r} = \int_0^{2\pi} (a\cos t)(b\cos t)\, dt \\ &= ab \int_0^{2\pi} \cos^2 t\, dt \\ &= ab \int_0^{2\pi} \frac{1 + \cos 2t}{2}\, dt \\ &= \pi ab + \frac{ab}{4} \sin 2t \Big|_0^{2\pi} \\ &= \pi ab \end{aligned}$$

The ellipse is shown in Figure 18.6.

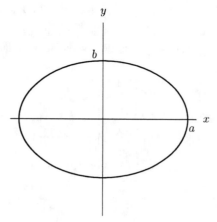

Figure 18.6: $\dfrac{x^2}{a^2} + \dfrac{y^2}{b^2} = 1$

Solutions for Section 18.5

1. We use the polar coordinates r and θ as parameters:

    $$x = -1 + r\cos\theta, \quad y = 2 + r\sin\theta, \qquad 2 \le r \le 3, 0 \le \theta \le 2\pi.$$

    To obtain the annulus, the values of r and θ vary on a rectangle.

5. We prove this by breaking each path into four pieces, as shown in Figure 18.5 on page 377 of the text. For example, the path C_1 is parameterized by

    $$\vec{r} = \vec{r}(s, c),$$

    so

    $$\int_{C_1} \vec{F} \cdot d\vec{r} = \int_a^b \vec{F} \cdot \frac{\partial \vec{r}}{\partial s}\, ds = \int_a^b G_1\, ds = \int_{D_1} \vec{G} \cdot d\vec{u}.$$

    The other pieces work the same way.

Solutions for Chapter 18 Review

1. (a) The line integral around A is negative, because the vectors of the field are all pointing in the opposite direction to the direction of the path.
    (b) Along C_1, the line integral is positive, since $\vec{F}$ points in the same direction as the curve. Along C_2 or C_4, the line integral is zero, since $\vec{F}$ is perpendicular to the curve everywhere. Along C_3, the line integral is negative, since $\vec{F}$ points in the opposite direction to the curve.
    (c) The line integral around C is negative because C_3 is longer than C_1 and the magnitude of the field is bigger along C_3 than C_1.

5.

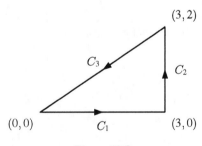

Figure 18.7

The triangle C consists of the three paths shown in Figure 18.7. Write $C = C_1 + C_2 + C_3$ where C_1, C_2, and C_3 are parameterized by

$$C_1 : (t, 0) \text{ for } 0 \le t \le 3; \quad C_2 : (3, t) \text{ for } 0 \le t \le 2; \quad C_3 : (3 - 3t, 2 - 2t) \text{ for } 0 \le t \le 1.$$

Then

$$\int_C \vec{F} \cdot d\vec{r} = \int_{C_1} \vec{F} \cdot d\vec{r} + \int_{C_2} \vec{F} \cdot d\vec{r} + \int_{C_3} \vec{F} \cdot d\vec{r}$$

where

$$\int_{C_1} \vec{F} \cdot d\vec{r} = \int_0^3 \vec{F}(t,0) \cdot \vec{i} \, dt = \int_0^3 (2t+4) dt = (t^2+4t)\big|_0^3 = 21$$

$$\int_{C_2} \vec{F} \cdot d\vec{r} = \int_0^2 \vec{F}(3,t) \cdot \vec{j} \, dt = \int_0^2 (5t+3) dt = (5t^2/2 + 3t)\big|_0^2 = 16$$

$$\int_{C_3} \vec{F} \cdot d\vec{r} = \int_0^1 \vec{F}(3-3t, 2-2t) \cdot (-3\vec{i} - 2\vec{j}) dt$$

$$= \int_0^1 ((-4t+8)\vec{i} + (-19t+13)\vec{j}) \cdot (-3\vec{i} - 2\vec{j}) dt$$

$$= 50 \int_0^1 (t-1) dt = -25.$$

So

$$\int_C \vec{F} \, d\vec{r} = 21 + 16 - 25 = 12.$$

9. True. You can trace out C_2 using the same subdivisions, but each $\Delta \vec{r}$ will have the opposite sign as before and will be traced out twice, so $\int_{C_2} \vec{F} \cdot d\vec{r} = -2 \int_{C_1} \vec{F} \cdot d\vec{r} = -6$.

13. (a) By the chain rule

$$\frac{dh}{dt} = \frac{\partial f}{\partial x} \frac{dx}{dt} + \frac{\partial f}{\partial y} \frac{dy}{dt} = f_x x'(t) + f_y y'(t),$$

which is the result we want.

(b) Using the parameterization of C that we were given,

$$\int_C \text{grad} f \cdot d\vec{r} = \int_a^b (f_x(x(t), y(t))\vec{i} + f_y(x(t), y(t))\vec{j}) \cdot (x'(t)\vec{i} + y'(t)\vec{j}) dt$$

$$= \int_a^b (f_x(x(t), y(t))x'(t) + f_y(x(t), y(t))y'(t)) dt.$$

Using the result of part (a), this gives us

$$\int_C \text{grad} f \cdot d\vec{r} = \int_a^b h'(t) dt$$
$$= h(b) - h(a) = f(Q) - f(P).$$

17. (a) An example of a central field is in Figure 18.8:

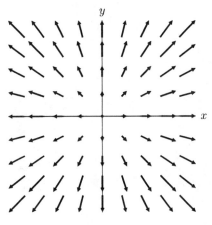

Figure 18.8

(b) The vectors of $\vec{F}$ are radial and the contours of f must be perpendicular to the vectors. Therefore, every contour must be a circle centered at the origin. Sketching some contours results in a diagram like that in Figure 18.9:

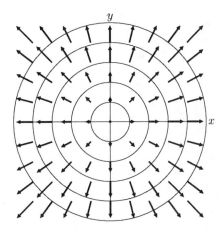

Figure 18.9

(c) No, not every gradient field is a central field, because there are gradient fields which are not perpendicular to circles. An example is the gradient of $f(x, y) = y$, where grad $f = \vec{j}$, so the gradient is parallel to the y axis. Thus, $\vec{F} = \vec{j}$ is an example of a gradient field which is not a central field.

(d) When a particle moves around a circle centered at O, no work is done, because $\vec{F}$ is tangent to the circle. Thus the only work done in moving from P to Q is in moving between the circles. Since $\vec{F}$ is central, the work done on any radial line between C_1 and C_3, for example, depends on only the radii of C_1 and C_3 ($\vec{F}$ is parallel to this path and its magnitude is a function of the distance to the center of the circle only). For that reason, on a path which goes from C_1 to C_2 and then from C_2 to C_3, the same amount of work will be done as on a path direct from C_1 to C_3.

(e) Pick any two points P and Q. Any path between them can be well-approximated by a path which is partly radial and partly around a circle centered at O. By the answer to part d), the work along any such path depends only on the radii of the circles on which P and Q sit, not on the path. Thus, the work done is independent of the path. Hence, $\vec{F}$ must be path-independent and therefore a gradient field.

CHAPTER NINETEEN

1. (a) The flux is positive, since $\vec{F}$ points in direction of positive x-axis, the same direction as the normal vector.
   (b) The flux is negative, since below the xy-plane $\vec{F}$ points towards negative x-axis, which is opposite the orientation of the surface.
   (c) The flux is zero. Since $\vec{F}$ has only an x-component, there is no flow across the surface.
   (d) The flux is zero. Since $\vec{F}$ has only an x-component, there is no flow across the surface.
   (e) The flux is zero. Since $\vec{F}$ has only an x-component, there is no flow across the surface.

5. (a) $\vec{v} \cdot \vec{A} = (2\vec{i} + 3\vec{j} + 5\vec{k}) \cdot \vec{k} = 5$.
   (b) $\vec{v} \cdot \vec{A} = (2\vec{i} + 3\vec{j} + 5\vec{k}) \cdot 2\vec{i} = 4$.
   (c) The rectangle lies in the plane $z + 2y = 2$. So a normal vector is $2\vec{j} + \vec{k}$ and a unit normal vector is $\frac{1}{\sqrt{5}}(2\vec{j} + \vec{k})$. Since this points in the positive z-direction it is indeed an orientation for the rectangle. Since the area of this rectangle is $\sqrt{5}$ we have $\vec{A} = 2\vec{j} + \vec{k}$,
   $\vec{v} \cdot \vec{A} = (2\vec{i} + 3\vec{j} + 5\vec{k}) \cdot (2\vec{j} + \vec{k}) = 6 + 5 = 11$.
   (d) The rectangle lies in the plane $z + 2x = 2$. So $2\vec{i} + \vec{k}$ is a normal vector and $\frac{1}{\sqrt{5}}(2\vec{i} + \vec{k})$ is a unit normal vector. Since this points in both the positive x-axis and the positive z it is an orientation for this surface. Since the area of the rectangle is $\sqrt{5}$, we have $\vec{A} = 2\vec{i} + \vec{k}$ and $\vec{v} \cdot \vec{A} = (2\vec{i} + 3\vec{j} + 5\vec{k}) \cdot (2\vec{i} + \vec{k}) = 4 + 5 = 9$.

9.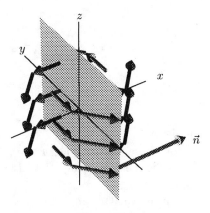

Figure 19.1

See Figure 19.1. The vector field is a vortex going around the z-axis, and the square is centered on the x-axis, so the flux going across one half of the square is balanced by the flux coming back across the other half. Thus, the net flux is zero, so

$$\int_S \vec{F} \cdot d\vec{A} = 0.$$

13. All the vectors in the vector field point horizontally (because their z-component is zero), and the surface is horizontal, so there is no flow through the surface and the flux is zero.

17. (a) Figure 19.2 shows the electric field $\vec{E}$. Note that $\vec{E}$ points radially outward from the z-axis.

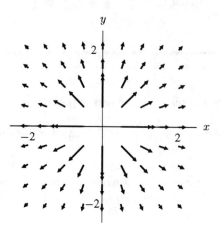

Figure 19.2: The electric field in the xy-plane due to a line of positive charge uniformly distributed along the z-axis:

$$\vec{E}(x, y, 0) = 2\lambda \frac{x\vec{i} + y\vec{j}}{x^2 + y^2}$$

(b) On the cylinder $x^2 + y^2 = R^2$, the electric field $\vec{E}$ points in the same direction as the outward normal $\vec{n}$, and

$$\|\vec{E}\| = \frac{2\lambda}{R^2} \|x\vec{i} + y\vec{j}\| = \frac{2\lambda}{R}.$$

So

$$\int_S \vec{E} \cdot d\vec{A} = \int_S \vec{E} \cdot \vec{n} \, dA = \int_S \|\vec{E}\| \, dA = \int_S \frac{2\lambda}{R} \, dA$$
$$= \frac{2\lambda}{R} \int_S dA = \frac{2\lambda}{R} \cdot \text{Area of } S = \frac{2\lambda}{R} \cdot 2\pi R h = 4\pi \lambda h,$$

which is positive, as we expected.

21. (a) If we examine the equation for $\vec{v}$, we see that when $r = 0$, that is, at the center of the pipe, $\vec{v}(0)$ becomes $u\vec{i}$. So u is the speed at the center of the pipe; it is also the maximum speed since $u(1 - r^2/a^2)$ reaches its maximum at $r = 0$.

(b) The flow rate at the wall of the pipe (where $r = a$) is

$$\vec{v}(a) = u(1 - a^2/a^2)\vec{i} = \vec{0}.$$

(c) To find the flux through a circular cross-sectional area, we use polar coordinates in the plane perpendicular to the velocity. In these coordinates, an infinitesimal area, $d\vec{A}$ becomes $r \, dr \, d\theta \vec{i}$. So the flux is given by

$$\text{Flux} = \int_S \vec{v} \cdot d\vec{A} = \int_S u(1 - r^2/a^2)\vec{i} \cdot r \, dr \, d\theta \vec{i} = \int_0^{2\pi} \int_0^a u(1 - r^2/a^2) r \, dr \, d\theta$$
$$= 2\pi u \int_0^a \left(r - \frac{r^3}{a^2} \right) dr = 2\pi u \left(\frac{a^2}{2} - \frac{a^2}{4} \right) = \frac{\pi u a^2}{2}.$$

Solutions for Section 19.2

1. Using $z = f(x, y) = x + y$, we have $d\vec{A} = (-\vec{i} - \vec{j} + \vec{k})\, dx\, dy$. As S is oriented upward, we have

$$\int_S \vec{F} \cdot d\vec{A} = \int_0^3 \int_0^2 ((x-y)\vec{i} + (x+y)\vec{j} + 3x\vec{k}) \cdot (-\vec{i} - \vec{j} + \vec{k})\, dxdy$$

$$= \int_0^3 \int_0^2 (-x + y - x - y + 3x)\, dxdy = \int_0^3 \int_0^2 x\, dxdy = 6.$$

5.

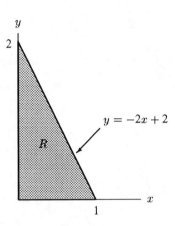

Figure 19.3

Writing the surface S as $z = f(x, y) = -2x - 4y + 1$, we have

$$d\vec{A} = (-f_x\vec{i} - f_y\vec{j} + \vec{k})dxdy.$$

With R as shown in Figure 19.3, we have

$$\int_S \vec{F} \cdot d\vec{A} = \int_R \vec{F}(x, y, f(x,y)) \cdot (-f_x\vec{i} - f_y\vec{j} + \vec{k})\, dxdy$$

$$= \int_R (3x\vec{i} + y\vec{j} + (-2x - 4y + 1)\vec{k}) \cdot (2\vec{i} + 4\vec{j} + \vec{k})\, dxdy$$

$$= \int_R (4x + 1)\, dxdy = \int_0^1 \int_0^{-2x+2} (4x + 1)\, dydx$$

$$= \int_0^1 (4x + 1)(-2x + 2)\, dx$$

$$= \int_0^1 (-8x^2 + 6x + 2)\, dx = \left(-\frac{8x^3}{3} + 3x^2 + 2x\right)\Big|_0^1 = \frac{7}{3}.$$

9. We have $0 \le z \le 6$ so $0 \le x^2 + y^2 \le 36$. Let R be the disk of radius 6 in the xy-plane centered at the origin. Because of the cone's point, the flux integral is improper; however, it does converge. We have

$$\int_S \vec{F} \cdot d\vec{A} = \int_R \vec{F}(x, y, f(x,y)) \cdot (-f_x\vec{i} - f_y\vec{j} + \vec{k})\, dxdy$$

$$= \int_R (-x\sqrt{x^2+y^2}\vec{i} - y\sqrt{x^2+y^2}\vec{j} + (x^2+y^2)\vec{k})$$

$$\cdot \left(-\frac{x}{\sqrt{x^2+y^2}}\vec{i} - \frac{y}{\sqrt{x^2+y^2}}\vec{j} + \vec{k} \right) dx\,dy$$

$$= \int_R 2(x^2+y^2)\,dx\,dy$$

$$= 2\int_0^6 \int_0^{2\pi} r^3\,d\theta\,dr$$

$$= 4\pi \int_0^6 r^3\,dr = 1296\pi.$$

13. Since the radius of the cylinder is 1, using cylindrical coordinates we have

$$d\vec{A} = (\cos\theta\vec{i} + \sin\theta\vec{j})\,d\theta\,dz.$$

Thus,

$$\int_S \vec{F}\cdot d\vec{A} = \int_0^6 \int_0^{2\pi} (\cos\theta\vec{i} + \sin\theta\vec{j}) \cdot (\cos\theta\vec{i} + \sin\theta\vec{j})\,d\theta\,dz$$

$$= \int_0^6 \int_0^{2\pi} 1\,d\theta\,dz = 12\pi.$$

17. (a) The position vector of a point (R, θ, z) (in cylindrical coordinates) on the cylinder is given by

$$\vec{r} = R\cos\theta\vec{i} + R\sin\theta\vec{j} + z\vec{k}.$$

So $\|\vec{r}\| = \sqrt{R^2 + z^2}$. Furthermore, for an area element on the cylinder we have the following

$$d\vec{A} = (\cos\theta\vec{i} + \sin\theta\vec{j})R\,dz\,d\theta,$$

and the integral is:

$$\int_S \vec{E}\cdot d\vec{A} = \int_0^{2\pi} \int_{-H}^{H} q\frac{\vec{r}}{\|\vec{r}\|^3} \cdot (\cos\theta\vec{i} + \sin\theta\vec{j})R\,dz\,d\theta$$

$$= q\int_0^{2\pi} \int_{-H}^{H} \frac{R\cos^2\theta + R\sin^2\theta}{(R^2+z^2)^{3/2}} R\,dz\,d\theta = 2\pi q \int_{-H}^{H} \frac{R^2\,dz}{(R^2+z^2)^{3/2}}.$$

In order to compute this one variable integral, we write:

$$\int_{-H}^{H} \frac{R^2\,dz}{(R^2+z^2)^{3/2}} = \int_{-H}^{H} \frac{(R^2+z^2)\,dz}{(R^2+z^2)^{3/2}} - \int_{-H}^{H} \frac{z^2\,dz}{(R^2+z^2)^{3/2}}$$

$$= \int_{-H}^{H} \frac{dz}{(R^2+z^2)^{1/2}} - \left(\int_{-H}^{H} \frac{dz}{(R^2+z^2)^{1/2}} - \frac{z}{(R^2+z^2)}\bigg|_{-H}^{H} \right)$$

$$= \int_{-H}^{H} \frac{dz}{(R^2+z^2)^{1/2}} - \int_{-H}^{H} \frac{dz}{(R^2+z^2)^{1/2}} + \frac{z}{(R^2+z^2)}\bigg|_{-H}^{H}$$

$$= \frac{2H}{\sqrt{R^2+H^2}}.$$

(The integral $\displaystyle\int_{-H}^{H} \frac{z^2 dz}{(R^2 + z^2)^{3/2}} = \int_{-H}^{H} z\frac{z dz}{(R^2 + z^2)^{3/2}}$ was computed using integration by parts).

Therefore

$$\int_S \vec{E} \cdot d\vec{A} = 4\pi q\frac{H}{\sqrt{R^2 + H^2}}.$$

(b) (i) Let R be fixed. We have

$$\lim_{H \to 0} \int_S \vec{E} \cdot d\vec{A} = \lim_{H \to 0} 4\pi q\frac{H}{\sqrt{H^2 + R^2}} = 0.$$

$$\lim_{H \to \infty} \int_S \vec{E} \cdot d\vec{A} = \lim_{H \to \infty} 4\pi q\frac{H}{\sqrt{H^2 + R^2}} = 4\pi q.$$

(ii) Now let H be fixed. We have

$$\lim_{R \to 0} \int_S \vec{E} \cdot d\vec{A} = \lim_{R \to 0} 4\pi q\frac{H}{\sqrt{H^2 + R^2}} = 4\pi q.$$

$$\lim_{R \to \infty} \int_S \vec{E} \cdot d\vec{A} = \lim_{R \to \infty} 4\pi q\frac{H}{\sqrt{H^2 + R^2}} = 0.$$

Solutions for Section 19.3

1. Since S is given by

$$\vec{r}(s, t) = (s + t)\vec{i} + (s - t)\vec{j} + (s^2 + t^2)\vec{k},$$

we have

$$\frac{\partial \vec{r}}{\partial s} = \vec{i} + \vec{j} + 2s\vec{k} \quad \text{and} \quad \frac{\partial \vec{r}}{\partial t} = \vec{i} - \vec{j} + 2t\vec{k},$$

and

$$\frac{\partial \vec{r}}{\partial s} \times \frac{\partial \vec{r}}{\partial t} = \begin{vmatrix} \vec{i} & \vec{j} & \vec{k} \\ 1 & 1 & 2s \\ 1 & -1 & 2t \end{vmatrix} = (2s + 2t)\vec{i} + (2s - 2t)\vec{j} - 2\vec{k}.$$

Since the $\vec{i}$ component of this vector is positive for $0 < s < 1, 0 < t < 1$, it points away from the z-axis, and so has the opposite orientation to the one specified. Thus, we use

$$d\vec{A} = -\frac{\partial \vec{r}}{\partial s} \times \frac{\partial \vec{r}}{\partial t} \, ds \, dt,$$

and so we have

$$\int_S \vec{F} \cdot d\vec{A} = -\int_0^1 \int_0^1 (s^2 + t^2)\vec{k} \cdot \left((2s + 2t)\vec{i} + (2s - 2t)\vec{j} - 2\vec{k}\right) ds \, dt$$

$$= 2\int_0^1 \int_0^1 (s^2 + t^2) \, ds \, dt = 2\int_0^1 \left(\frac{s^3}{3} + st^2\right)\Big|_{s=0}^{s=1} dt$$

$$= 2\int_0^1 \left(\frac{1}{3} + t^2\right) dt = 2\left(\frac{1}{3}t + \frac{t^3}{3}\right)\Big|_0^1 = 2\left(\frac{1}{3} + \frac{1}{3}\right) = \frac{4}{3}.$$

5.   Using cylindrical coordinates, we see that the surface S is parameterized by

$$\vec{r}(r, \theta) = r \cos \theta \vec{i} + r \sin \theta \vec{j} + r \vec{k}.$$

We have

$$\frac{\partial \vec{r}}{\partial r} \times \frac{\partial \vec{r}}{\partial \theta} = \begin{vmatrix} \vec{i} & \vec{j} & \vec{k} \\ \cos \theta & \sin \theta & 1 \\ -r \sin \theta & r \cos \theta & 0 \end{vmatrix} = -r \cos \theta \vec{i} - r \sin \theta \vec{j} + r \vec{k}.$$

Since the vector $\partial \vec{r}/\partial r \times \partial \vec{r}/\partial \theta$ points upward, in the direction opposite to the specified orientation, we use $d\vec{A} = -\left(\partial \vec{r}/\partial r \times \partial \vec{r}/\partial \theta\right) dr \, d\theta$. Hence

$$
\begin{aligned}
\int_S \vec{F} \cdot d\vec{A} &= -\int_0^{2\pi} \int_0^R (r^5 \cos^2 \theta \sin^2 \theta \vec{k}) \cdot (-r \cos \theta \vec{i} - r \sin \theta \vec{j} + r \vec{k}) \, dr \, d\theta \\
&= -\int_0^{2\pi} \int_0^R r^6 \cos^2 \theta \sin^2 \theta \, dr \, d\theta \\
&= -\frac{R^7}{7} \int_0^{2\pi} \sin^2 \theta \cos^2 \theta \, d\theta \\
&= -\frac{R^7}{7} \int_0^{2\pi} \sin^2 \theta (1 - \sin^2 \theta) \, d\theta \\
&= -\frac{R^7}{7} \int_0^{2\pi} (\sin^2 \theta - \sin^4 \theta) \, d\theta \\
&= -\left(\frac{R^7}{7}\right)\left(\frac{\pi}{4}\right) = \frac{-\pi}{28} R^7.
\end{aligned}
$$

The cone is not differentiable at the point $(0,0)$. However the flux integral, which is improper, converges.

9.   If S is the part of the graph of $z = f(x, y)$ lying over a region R in the xy-plane, then S is parameterized by

$$\vec{r}(x, y) = x\vec{i} + y\vec{j} + f(x, y)\vec{k}, \qquad (x, y) \text{ in } R.$$

So

$$\frac{\partial \vec{r}}{\partial x} \times \frac{\partial \vec{r}}{\partial y} = (\vec{i} + f_x \vec{k}) \times (\vec{j} + f_y \vec{k}) = -f_x \vec{i} - f_y \vec{j} + \vec{k}.$$

Since the $\vec{k}$ component is positive, this points upward, so if S is oriented upward

$$d\vec{A} = (-f_x \vec{i} - f_y \vec{j} + \vec{k}) \, dx \, dy$$

and therefore we have the expression for the flux integral obtained on page 399:

$$\int_S \vec{F} \cdot d\vec{A} = \int_R \vec{F}(x, y, f(x, y)) \cdot (-f_x \vec{i} - f_y \vec{k} + \vec{k}) \, dx \, dy.$$

Solutions for Chapter 19 Review━━━━

1.   For convention, orient the square so that the positive direction of flow is from down to up. Then when S is far up the positive z-axis, the flux is positive and large, because $\|\vec{r}\|$ is large. As S moves down, the flux gets

smaller. When S reaches the xy-plane, the flux is zero, because $\vec{r}$ and $\Delta\vec{A}$ are perpendicular. As S moves down more, the flux becomes more and more negative.

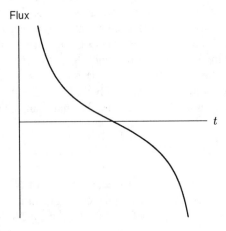

Figure 19.4

5. The flux through the surface equals

$$\vec{v} \cdot \vec{A}$$

where $\vec{A}$ is the area vector of the surface. The vector $\vec{i}$ is normal to the yz-plane. The area of the triangle is equal to 4, so the area vector of the triangle equals

$$\vec{A} = 4\vec{i}.$$

Thus

$$\vec{v} \cdot \vec{A} = 4$$

9. We have $d\vec{A} = \vec{k}\, dA$, so

$$\int_S \vec{F} \cdot d\vec{A} = \int_S (z\vec{i} + y\vec{j} + 2x\vec{k}) \cdot \vec{k}\, dA = \int_S 2x\, dA$$
$$= \int_0^3 \int_0^2 2x\, dx\, dy = 12.$$

13. There is no flux through the base or top of the cylinder because the vector field is parallel to these faces. For the curved surface, consider a small patch with area $\Delta\vec{A}$. The vector field is pointing radially outward from the z-axis and so is parallel to $\Delta\vec{A}$. Since $\|\vec{F}\| = \sqrt{x^2 + y^2} = 2$ on the curved surface of the cylinder, we have $\vec{F} \cdot \Delta\vec{A} = \|\vec{F}\|\|\Delta\vec{A}\| = 2\Delta A$. Replacing ΔA with dA, we get

$$\int_S \vec{F} \cdot d\vec{A} = \int_{\substack{\text{Curved} \\ \text{surface}}} 2\, dA = 2(\text{Area of curved surface}) = 2(2\pi \cdot 2 \cdot 3) = 24\pi.$$

17. (a) Consider two opposite faces of the cube, S_1 and S_2. The corresponding area vectors are $\vec{A}_1 = 4\vec{i}$ and $\vec{A}_2 = -4\vec{i}$ (since the side of the cube has length 2). Since $\vec{E}$ is constant, we find the flux by taking the dot product, giving

$$\text{Flux through } S_1 = \vec{E} \cdot \vec{A}_1 = (a\vec{i} + b\vec{j} + c\vec{k}) \cdot 4\vec{i} = 4a.$$

Flux through $S_2 = \vec{E} \cdot \vec{A}_2 = (a\vec{i} + b\vec{j} + c\vec{k}) \cdot (-4\vec{i}) = -4a$.

Thus the fluxes through S_1 and S_2 cancel. Arguing similarly, we conclude that, for any pair of opposite faces, the sum of the fluxes of $\vec{E}$ through these faces is zero. Hence, by addition, $\int_S \vec{E} \cdot d\vec{A} = 0$.

(b) The basic idea is the same as in part (a), except that we now need to use Riemann sums. First divide S into two hemispheres H_1 and H_2 by the equator C located in a plane perpendicular to $\vec{E}$. For a tiny patch S_1 in the hemisphere H_1, consider the patch S_2 in the opposite hemisphere which is symmetric to S_1 with respect to the center O of the sphere. The area vectors $\Delta \vec{A}_1$ and $\Delta \vec{A}_2$ satisfy $\Delta \vec{A}_2 = -\Delta \vec{A}_1$, so if we consider S_1 and S_2 to be approximately flat, then $\vec{E} \cdot \Delta \vec{A}_1 = -\vec{E} \cdot \Delta \vec{A}_2$. By decomposing H_1 and H_2 into small patches as above and using Riemann sums, we get

$$\int_{H_1} \vec{E} \cdot d\vec{A} = -\int_{H_2} \vec{E} \cdot d\vec{A}, \quad \text{so} \quad \int_S \vec{E} \cdot d\vec{A} = 0.$$

(c) The reasoning in part (b) can be used to prove that the flux of $\vec{E}$ through any surface with a center of symmetry is zero. For instance, in the case of the cylinder, cut it in half with a plane $z = 1$ and denote the two halves by H_1 and H_2. Just as before, take patches in H_1 and H_2 with $\Delta A_1 = -\Delta A_2$, so that $\vec{E} \cdot \Delta A_1 = -\vec{E} \cdot \Delta \vec{A}_2$. Thus, we get

$$\int_{H_1} \vec{E} \cdot d\vec{A} = -\int_{H_2} \vec{E} \cdot d\vec{A},$$

which shows that

$$\int_S \vec{E} \cdot d\vec{A} = 0.$$

CHAPTER TWENTY

Solutions for Section 20.1

1. Two vector fields that have positive divergence everywhere are as follows:

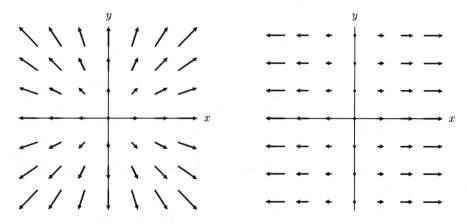

Figure 20.1   **Figure 20.2**

5. $\operatorname{div} \vec{F} = \dfrac{\partial}{\partial x}(-y) + \dfrac{\partial}{\partial y}(x) = 0$

9. In coordinates, we have

$$\vec{F}(x,y,z) = \frac{(x-x_0)}{\sqrt{(x-x_0)^2 + (y-y_0)^2 + (z-z_0)^2}}\vec{i} + \frac{(y-y_0)}{\sqrt{(x-x_0)^2 + (y-y_0)^2 + (z-z_0)^2}}\vec{j}$$

$$+ \frac{(z-z_0)}{\sqrt{(x-x_0)^2 + (y-y_0)^2 + (z-z_0)^2}}\vec{k}\,.$$

So if $(x,y,z) \neq (x_0, y_0, z_0)$, then

$$\begin{aligned}
\operatorname{div} \vec{F} &= \left(\frac{1}{\sqrt{(x-x_0)^2 + (y-y_0)^2 + (z-z_0)^2}} - \frac{(x-x_0)^2}{((x-x_0)^2 + (y-y_0)^2 + (z-z_0)^2)^{3/2}} \right) \\
&\quad + \left(\frac{1}{\sqrt{(x-x_0)^2 + (y-y_0)^2 + (z-z_0)^2}} - \frac{(y-y_0)^2}{((x-x_0)^2 + (y-y_0)^2 + (z-z_0)^2)^{3/2}} \right) \\
&\quad + \left(\frac{1}{\sqrt{(x-x_0)^2 + (y-y_0)^2 + (z-z_0)^2}} - \frac{(z-z_0)^2}{((x-x_0)^2 + (y-y_0)^2 + (z-z_0)^2)^{3/2}} \right) \\
&= \left(\frac{(x-x_0)^2 + (y-y_0)^2 + (z-z_0)^2}{((x-x_0)^2 + (y-y_0)^2 + (z-z_0)^2)^{3/2}} - \frac{(x-x_0)^2}{((x-x_0)^2 + (y-y_0)^2 + (z-z_0)^2)^{3/2}} \right) \\
&\quad + \left(\frac{(x-x_0)^2 + (y-y_0)^2 + (z-z_0)^2}{((x-x_0)^2 + (y-y_0)^2 + (z-z_0)^2)^{3/2}} - \frac{(y-y_0)^2}{((x-x_0)^2 + (y-y_0)^2 + (z-z_0)^2)^{3/2}} \right)
\end{aligned}$$

$$+ \left(\frac{(x - x_0)^2 + (y - y_0)^2 + (z - z_0)^2}{((x - x_0)^2 + (y - y_0)^2 + (z - z_0)^2)^{3/2}} - \frac{(z - z_0)^2}{((x - x_0)^2 + (y - y_0)^2 + (z - z_0)^2)^{3/2}} \right)$$

$$= \frac{3((x - x_0)^2 + (y - y_0)^2 + (z - z_0)^2) - ((x - x_0)^2 + (y - y_0)^2 + (z - z_0)^2)}{((x - x_0)^2 + (y - y_0)^2 + (z - z_0)^2)^{3/2}}$$

$$= \frac{2}{\sqrt{(x - x_0)^2 + (y - y_0)^2 + (z - z_0)^2}} = \frac{2}{\|\vec{r} - \vec{r}_0\|}.$$

13. Using $\text{div}(g\vec{F}) = (\text{grad } g) \cdot \vec{F} + g \, \text{div } \vec{F}$, we have

$$\text{div } \vec{F} = \frac{1}{\|\vec{r}\|^p} \text{div}(\vec{a} \times \vec{r}) + \text{grad}(\frac{1}{\|\vec{r}\|^p}) \cdot \vec{a} \times \vec{r}$$

$$= \frac{1}{\|\vec{r}\|^p} 0 + \frac{-p}{\|\vec{r}\|^{p+2}} \vec{r} \cdot (\vec{a} \times \vec{r})$$

$$= 0 \quad \text{since } \vec{r} \text{ and } \vec{a} \times \vec{r} \text{ are perpendicular.}$$

17. (a) Positive. The inflow from the lower left is less than the outflow from the upper right. The net outflow is positive.
   (b) Zero. The inflow on the right side is equal to outflow on the left.
   (c) Negative. The inflow from above is greater than the outflow below. The net outflow is negative.

21. (a) Since $2\vec{i} + 3\vec{k}$ is a constant field, its contribution to the flux is zero (flux in cancels flux out). Therefore $\int \vec{F} \cdot d\vec{A} = \int (y\vec{j}) \cdot d\vec{A} = \int_{S_3} y\vec{j} \cdot d\vec{A} + \int_{S_4} y\vec{j} \cdot d\vec{A}$ since only S_3 and S_4 are perpendicular to $y\vec{j}$. On S_3, $y = 0$ so $\int_{S_3} y\vec{j} \cdot d\vec{A} = 0$. On S_4, $y = c$ and normal is in the positive y-direction, so $\int_{S_4} y\vec{j} \cdot d\vec{A} = c(\text{Area of } S_4) = c \cdot c^2 = c^3$. Thus, total flux $= c^3$.
   (b) Using the geometric definition of convergence

$$\text{div } \vec{F} = \lim_{c \to 0} \left(\frac{\text{Flux through box}}{\text{Volume of box}} \right)$$

$$= \lim_{c \to 0} \left(\frac{c^3}{c^3} \right) = 1.$$

   (c)

$$\frac{\partial}{\partial x}(2) + \frac{\partial}{\partial y}(y) + \frac{\partial}{\partial z}(3) = 0 + 1 + 0 = 1.$$

25. (a) Translating the vector field into rectangular coordinates gives, if $(x, y, z) \neq (0, 0, 0)$

$$\vec{E}(x, y, z) = \frac{kx}{(x^2 + y^2 + z^2)^{3/2}} \vec{i} + \frac{ky}{(x^2 + y^2 + z^2)^{3/2}} \vec{j} + \frac{kz}{(x^2 + y^2 + z^2)^{3/2}} \vec{k}.$$

We now take the divergence of this to get

$$\text{div } \vec{E} = k \left(-3 \frac{x^2 + y^2 + z^2}{(x^2 + y^2 + z^2)^{5/2}} + \frac{3}{(x^2 + y^2 + z^2)^{3/2}} \right)$$

$$= 0.$$

(b)   Let S be the surface of a sphere centered at the origin. We have seen that for this field, the flux $\int \vec{E} \cdot d\vec{A}$ is the same for all such spheres, regardless of their radii. So let the constant c stand for $\int \vec{E} \cdot d\vec{A}$. Then

$$\text{div } \vec{E}\,(0,0,0) = \lim_{\text{vol}\to 0} \frac{\int \vec{E} \cdot d\vec{A}}{\text{Volume inside } S} = \lim_{\text{vol}\to 0} \frac{c}{\text{Volume}}.$$

(c)   For a point charge, the charge density is not defined. The charge density is 0 everywhere else.

29.   (a)   At any point $\vec{r} = x\vec{i} + y\vec{j}$, the direction of the vector field $\vec{v}$ is pointing away from the origin, which means it is of the form $\vec{v} = f\vec{r}$ for some positive function f, whose value can vary depending on $\vec{r}$. The magnitude of $\vec{v}$ depends only on the distance r, thus f must be a function depending only on r, which is equivalent to depending only on r^2 since $r \geq 0$. So $\vec{v} = f(r^2)\vec{r} = \left(f(x^2 + y^2)\right)(x\vec{i} + y\vec{j})$.

(b)   At $(x,y) \neq (0,0)$ the divergence of $\vec{v}$ is

$$\text{div } \vec{v} = \frac{\partial(K(x^2 + y^2)^{-1}x)}{\partial x} + \frac{\partial(K(x^2 + y^2)^{-1}y)}{\partial y} = \frac{Ky^2 - Kx^2}{(x^2 + y^2)^2} + \frac{Kx^2 - Ky^2}{(x^2 + y^2)^2} = 0.$$

Therefore, $\vec{v}$ is a point source at the origin.

(c)   The magnitude of $\vec{v}$ is

$$\|\vec{v}\| = K(x^2 + y^2)^{-1}|x\vec{i} + y\vec{j}| = K(x^2 + y^2)^{-1}(x^2 + y^2)^{1/2} = K(x^2 + y^2)^{-1/2} = \frac{K}{r}.$$

(d)   The vector field looks like that in Figure 20.3:

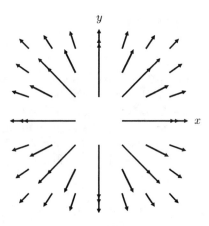

Figure 20.3

(e)   We need to show that grad $\phi = \vec{v}$.

$$\text{grad }\phi = \frac{\partial}{\partial x}\left(\frac{K}{2}\log(x^2 + y^2)\right)\vec{i} + \frac{\partial}{\partial y}\left(\frac{K}{2}\log(x^2 + y^2)\right)\vec{j}$$
$$= \frac{Kx}{x^2 + y^2}\vec{i} + \frac{Ky}{x^2 + y^2}\vec{j}$$
$$= K(x^2 + y^2)^{-1}(x\vec{i} + y\vec{j})$$
$$= \vec{v}$$

Solutions for Section 20.2

1. First directly: On the faces $x = 0, y = 0, z = 0$, the flux is zero. On the face $x = 2$, a unit normal is $\vec{i}$ and $d\vec{A} = dA\vec{i}$. So

$$\int_{S_{x=2}} \vec{r} \cdot d\vec{A} = \int_{S_{x=2}} (2\vec{i} + y\vec{j} + z\vec{k}) \cdot (dA\vec{i})$$

(since on that face, $x = 2$)

$$= \int_{S_{x=2}} 2 dA = 2 \cdot (\text{Area of face}) = 2 \cdot 4 = 8.$$

In exactly the same way, you get

$$\int_{S_{y=2}} \vec{r} \cdot d\vec{A} = \int_{S_{z=2}} \vec{r} \cdot d\vec{A} = 8,$$

so

$$\int_{S} \vec{r} \cdot d\vec{A} = 3 \cdot 8 = 24.$$

Now using divergence:

$$\operatorname{div} \vec{F} = \frac{\partial x}{\partial x} + \frac{\partial y}{\partial y} + \frac{\partial z}{\partial z} = 3,$$

so

$$\text{Flux} = \int_0^2 \int_0^2 \int_0^2 3 \, dx \, dy \, dz = 3 \cdot (\text{Volume of Cube}) = 3 \cdot 8 = 24$$

5. Finding flux directly:
    1) On bottom face, $z = 0$ so $\vec{F} = x^2\vec{i} + 2y^2\vec{j}$ is parallel to face so flux is zero.
    2) On front face, $y = 0$ so $\vec{F} = x^2\vec{i} + 3z^2\vec{k}$ is parallel to face so flux is zero.
    3) On back face, $y = 1$ so $\vec{F} = x^2\vec{i} + 2\vec{j} + 3z^2\vec{k}$ and $\vec{A} = \vec{j}$ so flux is 2.
    4) On top face, $z = 1$ so $\vec{F} = x^2\vec{i} + 2y^2\vec{j} + 3\vec{k}$ and $\vec{A} = \vec{k}$ so flux is 3.
    5) On side $x = 1$, $\vec{F} = \vec{i} + 2y^2\vec{j} + 3z^2\vec{k}$ and $\vec{A} = -\vec{i}$ so flux is -1.
    6) On side $x = 2$, $\vec{F} = 4\vec{i} + 2y^2\vec{j} + 3z^2\vec{k}$ and $\vec{A} = \vec{i}$ so flux is 4.
    Total flux is thus 8.
       By the Divergence Theorem:

$$\operatorname{div} \vec{F} = 2x + 4y + 6z$$

So

$$\int_S \vec{F} \cdot d\vec{A} = \int_V (2x + 4y + 6z) dV = 2 \int_1^2 \int_0^1 \int_0^1 (x + 2y + 3z) dz \, dy \, dx$$

$$= 2 \int_1^2 \int_0^1 \left[xz + 2yz + \frac{3z^2}{2} \right]_0^1 dy \, dx = 2 \int_1^2 \int_0^1 \left(x + 2y + \frac{3}{2} \right) dy \, dx$$

$$= 2 \int_1^2 \left[xy + y^2 + \frac{3y}{2} \right]_0^1 dx = 2 \int_1^2 \left(x + 1 + \frac{3}{2} \right) dx$$

$$= \int_1^2 (2x + 5) dx = (x^2 + 5x) \Big|_1^2 = 8$$

9. (a) Taking partial derivatives, we have for $\vec{r} \neq \vec{0}$,

$$\text{div } \vec{F} = \text{div } \left(\frac{x\vec{i} + y\vec{j} + z\vec{k}}{(x^2 + y^2 + z^2)^{3/2}} \right)$$

$$= -\frac{3(x^2 + y^2 + z^2)}{(x^2 + y^2 + z^2)^{5/2}} + \frac{3}{(x^2 + y^2 + z^2)^{3/2}}$$

$$= 0.$$

(b) We could compute the flux out of the box S by computing the flux out of each side separately. However, since div $\vec{F} = \vec{0}$ everywhere except the origin, we instead consider a region W between the box S and a sphere S_b of radius b centered at the origin and which fits inside the box. If the sphere is oriented inward, since div $\vec{F} = \vec{0}$ throughout W, the Divergence Theorem says

$$0 = \int_W \text{div } \vec{F} \, dV = \int_S \vec{F} \cdot d\vec{A} + \int_{S_b} \vec{F} \cdot d\vec{A}.$$

The flux of $\vec{F}$ through S_b is easier to calculate than the flux through the box. Since S_b is oriented inward,

$$\int_{S_b} \vec{F} \cdot d\vec{A} = -\int_{S_b} \|\vec{F}\| \|d\vec{A}\| = -\int_{S_b} \frac{1}{b^2} \|dA\|$$

$$= -\frac{1}{b^2} \cdot \text{Surface area of } S_b = -\frac{1}{b^2} \cdot 4\pi b^2 = -4\pi.$$

Thus,

$$\int_S \vec{F} \cdot d\vec{A} = -\int_{S_b} \vec{F} \cdot d\vec{A} = 4\pi.$$

13. Since the divergence is zero at all points not containing the charge, the flux must be zero through any closed surface containing no charge. We imagine a surface composed of two concentric cylinders and their end-caps, where the axis of both cylinders is the z-axis. Then, since no charge is contained in the region enclosed, the flux through the surface must be zero. Now, we know that the field points away from the axis, which means it is parallel to the end-caps. Consequently, there must be no flux through the end-caps. This implies that the flux through the inner cylinder must equal the flux out of the outer cylinder. Since the strength of the field only depends upon the distance from the z axis, the flux through each cylinder is a constant. This implies that the following equation must hold

$$\text{Flux through each cylinder} = E_a 2\pi r_a L = E_b 2\pi r_b L$$

where E_a and E_b are the strengths of the field at r_a and r_b respectively, and L is the length of the cylinders. Dividing through, we can arrive at the following relationship:

$$E_a / E_b = r_b / r_a$$

If we take E_b to be a constant at a fixed value of r_b, then the equation can be simplified to

$$E_a = k / r_a$$

where $k = E_b r_b$. Thus we see that the strength of the field is proportional to $1/r$.

17. Suppose $\phi(x, y, z) = ax + by + cz + d$ is linear (a, b, c, d are constants). We have

$$\nabla^2 \phi(x, y, z) = \frac{\partial^2}{\partial x^2}(ax + by + cz + d) + \frac{\partial^2}{\partial y^2}(ax + by + cz + d)$$
$$+ \frac{\partial^2}{\partial z^2}(ax + by + cz + d) = 0$$

Hence ϕ is harmonic.

21. Let ϕ be our nonconstant harmonic function on the region R. Denote $\psi = -\phi$. Then ψ is nonconstant, and

$$\nabla^2 \psi = -\nabla^2 \phi = 0.$$

Hence, ψ is also harmonic.

Moreover, any minimum value for ϕ in R is a maximum value for ψ in the same region. Now use Example 5 on page 424 to derive that if ϕ has a minimum value at a point, ψ has a maximum one at the same point. So the point must be on the boundary of R.

25. Apply Divergence Theorem to the integral:

$$\int_S u \operatorname{grad} v \cdot d\vec{A} = \int_R \operatorname{div}(u \operatorname{grad} v)\, dV.$$

Now:

$$\operatorname{div}(u \operatorname{grad} v) = \operatorname{div}(u\frac{\partial v}{\partial x}\vec{i} + u\frac{\partial v}{\partial y}\vec{j} + u\frac{\partial v}{\partial z}\vec{k})$$
$$= \frac{\partial}{\partial x}(u\frac{\partial v}{\partial x}) + \frac{\partial}{\partial y}(u\frac{\partial v}{\partial y}) + \frac{\partial}{\partial z}(u\frac{\partial v}{\partial z})$$
$$= \frac{\partial u}{\partial x}\frac{\partial v}{\partial x} + \frac{\partial u}{\partial y}\frac{\partial v}{\partial y} + \frac{\partial u}{\partial z}\frac{\partial v}{\partial z} + u \cdot (\frac{\partial^2 v}{\partial x^2} + \frac{\partial^2 v}{\partial y^2} + \frac{\partial^2 v}{\partial z^2})$$
$$= \operatorname{grad} u \cdot \operatorname{grad} v.$$

Hence

$$\int_S u \operatorname{grad} v \cdot d\vec{A} = \int_R (\operatorname{grad} u \cdot \operatorname{grad} v)\, dV.$$

Similarly,

$$\int_S v \operatorname{grad} v \cdot d\vec{A} = \int_R \operatorname{div}(v \operatorname{grad} u)\, dV = \int_R (\operatorname{grad} v \cdot \operatorname{grad} u)\, dV.$$

Hence

$$\int_S u \operatorname{grad} v \cdot d\vec{A} = \int_S v \operatorname{grad} u \cdot d\vec{A}.$$

Solutions for Section 20.3

1. Using the definition in Cartesian coordinates, we have

$$\operatorname{curl} \vec{F} = \begin{vmatrix} \vec{i} & \vec{j} & \vec{k} \\ \frac{\partial}{\partial x} & \frac{\partial}{\partial y} & \frac{\partial}{\partial z} \\ x^2 - y^2 & 2xy & 0 \end{vmatrix}$$
$$= \left(\frac{\partial}{\partial y}(0) - \frac{\partial}{\partial z}(2xy)\right)\vec{i} + \left(-\frac{\partial}{\partial x}(0) + \frac{\partial}{\partial z}(x^2 - y^2)\right)\vec{j} + \left(\frac{\partial}{\partial x}(2xy) - \frac{\partial}{\partial y}(x^2 - y^2)\right)\vec{k}$$
$$= 4y\vec{k}.$$

5. Using the definition of Cartesian coordinates,

$$\text{curl}\,\vec{F} = \begin{vmatrix} \vec{i} & \vec{j} & \vec{k} \\ \frac{\partial}{\partial x} & \frac{\partial}{\partial y} & \frac{\partial}{\partial z} \\ 2yz & 3xz & 7xy \end{vmatrix}$$

$$= (7x - 3x)\vec{i} - (7y - 2y)\vec{j} + (3z - 2z)\vec{k}$$

$$= 4x\vec{i} - 5y\vec{j} + z\vec{k}\,.$$

9. The conjecture is that when the first component of $\vec{F}$ depends only on x, the second component depends only on y, and the third component depends only on z, that is, if

$$\vec{F} = F_1(x)\vec{i} + F_2(y)\vec{j} + F_3(z)\vec{k}$$

then

$$\text{curl}\,\vec{F} = \vec{0}$$

The reason for this is that if $\vec{F} = F_1(x)\vec{i} + F_2(y)\vec{j} + F_3(z)\vec{k}$, then

$$\text{curl}\,\vec{F} = \begin{vmatrix} \vec{i} & \vec{j} & \vec{k} \\ \frac{\partial}{\partial x} & \frac{\partial}{\partial y} & \frac{\partial}{\partial z} \\ F_1(x) & F_2(y) & F_3(z) \end{vmatrix}$$

$$= \left(\frac{\partial}{\partial y}F_3(z) - \frac{\partial}{\partial z}F_2(y)\right)\vec{i} + \left(-\frac{\partial}{\partial x}F_3(z) + \frac{\partial}{\partial z}F_1(x)\right)\vec{j} + \left(\frac{\partial}{\partial x}F_2(y) - \frac{\partial}{\partial y}F_1(x)\right)\vec{k}$$

$$= \vec{0}\,.$$

13. Let $\vec{C} = a\vec{i} + b\vec{j} + c\vec{k}$. Then

$$\text{curl}(\vec{F} + \vec{C}) = \left(\frac{\partial}{\partial y}(F_3 + c) - \frac{\partial}{\partial z}(F_2 + b)\right)\vec{i} + \left(\frac{\partial}{\partial z}(F_1 + a) - \frac{\partial}{\partial x}(F_3 + c)\right)\vec{j}$$

$$+ \left(\frac{\partial}{\partial x}(F_2 + b) - \frac{\partial}{\partial y}(F_1 + a)\right)\vec{k}$$

$$= \left(\frac{\partial F_3}{\partial y} - \frac{\partial F_2}{\partial z}\right)\vec{i} + \left(\frac{\partial F_1}{\partial z} - \frac{\partial F_3}{\partial x}\right)\vec{j} + \left(\frac{\partial F_2}{\partial x} - \frac{\partial F_1}{\partial y}\right)\vec{k}$$

$$= \text{curl}\,\vec{F}\,.$$

17.

$$\text{curl}(\phi\vec{F})$$

$$= \left(\frac{\partial}{\partial y}(\phi F_3) - \frac{\partial}{\partial z}(\phi F_2)\right)\vec{i} + \left(\frac{\partial}{\partial z}(\phi F_1) - \frac{\partial}{\partial x}(\phi F_3)\right)\vec{j} + \left(\frac{\partial}{\partial x}(\phi F_2) - \frac{\partial}{\partial y}(\phi F_1)\right)\vec{k}$$

$$= \left(\phi\frac{\partial F_3}{\partial y} + \frac{\partial\phi}{\partial y}F_3 - \phi\frac{\partial F_2}{\partial z} - \frac{\partial\phi}{\partial z}F_2\right)\vec{i} + \left(\phi\frac{\partial F_1}{\partial z} + \frac{\partial\phi}{\partial z}F_1 - \phi\frac{\partial F_3}{\partial x} - \frac{\partial\phi}{\partial x}F_3\right)\vec{j}$$

$$+ \left(\phi\frac{\partial F_2}{\partial x} + \frac{\partial \phi}{\partial x}F_2 - \phi\frac{\partial F_1}{\partial y} - \frac{\partial \phi}{\partial y}F_1\right)\vec{k}$$

$$= \phi\left(\left(\frac{\partial F_3}{\partial y} - \frac{\partial F_2}{\partial z}\right)\vec{i} + \left(\frac{\partial F_1}{\partial z} - \frac{\partial F_3}{\partial x}\right)\vec{j} + \left(\frac{\partial F_2}{\partial x} - \frac{\partial F_1}{\partial y}\right)\vec{k}\right)$$

$$+ \left(\left(\frac{\partial \phi}{\partial y}F_3 - \frac{\partial \phi}{\partial z}F_2\right)\vec{i} + \left(\frac{\partial \phi}{\partial z}F_1 - \frac{\partial \phi}{\partial x}F_3\right)\vec{j} + \left(\frac{\partial \phi}{\partial x}F_2 - \frac{\partial \phi}{\partial y}F_1\right)\vec{k}\right)$$

$$= \phi\,\text{curl}\,\vec{F} + \left(\frac{\partial \phi}{\partial x}\vec{i} + \frac{\partial \phi}{\partial y}\vec{j} + \frac{\partial \phi}{\partial z}\vec{k}\right) \times (F_1\vec{i} + F_2\vec{j} + F_3\vec{k})$$

$$= \phi\,\text{curl}\,\vec{F} + (\text{grad}\,\phi) \times \vec{F}.$$

21. C_2, C_3, C_4, C_6, since line integrals around C_1 and C_5 are clearly nonzero. You can see directly that $\int_{C_2} \vec{F} \cdot d\vec{r}$ and $\int_{C_6} \vec{F} \cdot d\vec{r}$ are zero, because C_2 and C_6 are perpendicular to their fields at every point.

25.   Let $\vec{v} = a\vec{i} + b\vec{j} + c\vec{k}$ and try

$$\vec{F} = \vec{v} \times \vec{r} = (a\vec{i} + b\vec{j} + c\vec{k}) \times (x\vec{i} + y\vec{j} + z\vec{k}) = (bz - cy)\vec{i} + (cx - az)\vec{j} + (ay - bx)\vec{k}.$$

Then

$$\text{curl}\,\vec{F} = \begin{vmatrix} \vec{i} & \vec{j} & \vec{k} \\ \frac{\partial}{\partial x} & \frac{\partial}{\partial y} & \frac{\partial}{\partial z} \\ bz - cy & cx - az & ay - bx \end{vmatrix} = 2a\vec{i} + 2b\vec{j} + 2c\vec{k}.$$

Taking $a = 1$, $b = -\frac{3}{2}$, $c = 2$ gives $\text{curl}\,\vec{F} = 2\vec{i} - 3\vec{j} + 4\vec{k}$, so the desired vector field is $\vec{F} = (-\frac{3}{2}z - 2y)\vec{i} + (2x - z)\vec{j} + (y + \frac{3}{2}x)\vec{k}$.

Solutions for Section 20.4

1.   No, because the curve C over which the integral is taken is not a closed curve, and so it is not the boundary of a surface.

5.   The circulation is the line integral $\int_C \vec{F} \cdot d\vec{r}$ which can be evaluated directly by parameterizing the circle, C. Or, since C is the boundary of a flat disk S, we can use Stokes' Theorem:

$$\int_C \vec{F} \cdot d\vec{r} = \int_S \text{curl}\,\vec{F} \cdot d\vec{A}$$

where S is the disk $x^2 + y^2 \le 1$, $z = 2$ and is oriented upward (using the right hand rule). Then $\text{curl}\,\vec{F} = -y\vec{i} - x\vec{j} + \vec{k}$ and the unit normal to S is $\vec{k}$. So

$$\int_S \text{curl}\,\vec{F} \cdot d\vec{A} = \int_S (-y\vec{i} - x\vec{j} + \vec{k}) \cdot \vec{k}\,dxdy$$

$$= \int_S 1\,dxdy$$

$$= \text{Area of } S$$

$$= \pi$$

9.   Use Stokes' theorem, applied to the surface R, oriented upwards. Since $\text{curl}\vec{F} = \vec{k}$ for $\vec{F} = \frac{1}{2}(-y\vec{i} + x\vec{j})$, we have $\frac{1}{2}\int_C(-y\vec{i} + x\vec{j}) \cdot d\vec{r} = \int_R \vec{k} \cdot d\vec{A} = \|\vec{k}\|(\text{area of } R) = \text{area of } R.$

13. (a) You can't say anything, because any surface bounded by the circle must intersect the z-axis. Since curl $\vec{F}$ is not defined on the z-axis, the surface integral in Stokes' Theorem is not defined.

(b) In this case curl $\vec{F}$ is defined and equal to 0 on a surface S bounded by the the circle, so Stokes' Theorem says that

$$\int_C \vec{F} \cdot d\vec{r} = \int_S \text{curl}\, \vec{F} \cdot d\vec{A} = 0.$$

17. Using Stokes' Theorem, the flux integral $\int_S \text{curl}\, \vec{F} \cdot d\vec{A}$ has the same value as the line integral $\int_C \vec{F} \cdot d\vec{r}$, where C is the boundary curve of S with the appropriate orientation. Here C is the unit circle $x^2 + y^2 = 1, z = 0$ oriented clockwise when viewed from above. We can parameterize C by $\vec{r}(t) = \cos t \vec{i} - \sin t \vec{j}$ with $0 \le t \le 2\pi$.

Then $\vec{r}'(t) = -\sin t \vec{i} - \cos t \vec{j}$, so

$$\int_C \vec{F} \cdot d\vec{r} = \int_0^{2\pi} (\sin t \vec{i} + \cos t \vec{j}) \cdot (-\sin t \vec{i} - \cos t \vec{j}) dt$$

$$= \int_0^{2\pi} (-\sin^2 t - \cos^2 t)\, dt$$

$$= \int_0^{2\pi} -1\, dt$$

$$= -2\pi$$

Solutions for Section 20.5

1. Since curl $\vec{F} = \vec{0}$ and $\vec{F}$ is defined everywhere, we know by the curl test that $\vec{F}$ is a gradient field. In fact, $\vec{F} = \text{grad}\, f$, where $f(x, y, z) = xyz + yz^2$, so f is a potential function for $\vec{F}$.

5. We must show curl $\vec{A} = \vec{B}$.

$$\text{curl}\, \vec{A} = \frac{\partial}{\partial y} \left(\frac{-I}{c} \ln(x^2 + y^2) \right) \vec{i} - \frac{\partial}{\partial x} \left(\frac{-I}{c} \ln(x^2 + y^2) \right) \vec{j}$$

$$= \frac{-I}{c} \left(\frac{2y}{x^2 + y^2} \right) \vec{i} + \frac{I}{c} \left(\frac{2x}{(x^2 + y^2)} \right) \vec{j}$$

$$= \frac{2I}{c} \left(\frac{-y\vec{i} + x\vec{j}}{x^2 + y^2} \right)$$

$$= \vec{B}.$$

9. (a) Yes. To show this, we use a version of the product rule for curl (Problem 17 on page 137):

$$\text{curl}(\phi\vec{F}) = \phi\, \text{curl}\, \vec{F} + (\text{grad}\, \phi) \times \vec{F},$$

where ϕ is a scalar function and $\vec{F}$ is a vector field. So

$$\text{curl}\left(q\frac{\vec{r}}{\|\vec{r}\|^3} \right) = \text{curl}\left(\frac{q}{\|\vec{r}\|^3}\vec{r} \right) = \frac{q}{\|\vec{r}\|^3}\, \text{curl}\, \vec{r} + \text{grad}\left(\frac{q}{\|\vec{r}\|^3} \right) \times \vec{r}$$

$$= \vec{0} + q\, \text{grad}\left(\frac{1}{\|\vec{r}\|^3} \right) \times \vec{r}$$

Since the level surfaces of $1/\|\vec{r}\|^3$ are spheres centered at the origin, $\mathrm{grad}(1/\|\vec{r}\|^3)$ is parallel to $\vec{r}$, so $\mathrm{grad}(1/\|\vec{r}\|^3) \times \vec{r} = \vec{0}$. Thus, $\mathrm{curl}\,\vec{E} = \vec{0}$.

(b) Yes. The domain of $\vec{E}$ is 3-space minus $(0,0,0)$. Any closed curve in this region is the boundary of a surface contained entirely in the region. (If the first surface you pick happens to contain $(0,0,0)$, change its shape slightly to avoid it.)

(c) Yes. Since $\vec{E}$ satisfies both conditions of the curl test, it must be a gradient field. In fact,

$$\vec{E} = \mathrm{grad}\left(-q\frac{1}{\|\vec{r}\|}\right).$$

13. (a) Since $\mathrm{curl}\,\mathrm{grad}\,\psi = 0$ for any function ψ, $\mathrm{curl}(\vec{A} + \mathrm{grad}\,\psi) = \mathrm{curl}\,\vec{A} + \mathrm{curl}\,\mathrm{grad}\,\psi = \mathrm{curl}\,\vec{A} = \vec{B}$.

(b) We have

$$\mathrm{div}(\vec{A} + \mathrm{grad}\,\psi) = \mathrm{div}\,\vec{A} + \mathrm{div}\,\mathrm{grad}\,\psi = \mathrm{div}\,\vec{A} + \nabla^2\psi.$$

Thus ψ should be chosen to satisfy the partial differential equation

$$\nabla^2\psi = -\,\mathrm{div}\,\vec{A}.$$

Solutions for Section 20.6

1. Suppose the cylinder goes from a to b along the z-axis, and has inner radius r_1 and outer radius r_2. The in cylindrical coordinates the region is parameterized by the rectangular region $a \le z \le b$, $r_1 \le r \le r_2$, and $0 \le \theta \le 2\pi$. The faces $\theta = 0$ and $\theta = 2\pi$ are pasted together.

5. (a) Consider a small parallelepiped with one corner at the point $\vec{r}_0 = (x_0, y_0, z_0)$ and with edges $\vec{a}\,\Delta x$, $\vec{b}\,\Delta y$, and $\vec{c}\,\Delta z$, as in Figure 20.4.

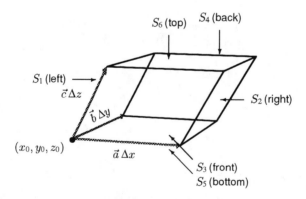

Figure 20.4

On S_1 (the left face of the parallelepiped shown in Figure 20.4) the outward area vector is $\Delta\vec{A} = -(\vec{b}\,\Delta y) \times (\vec{c}\,\Delta z) = -\vec{b} \times \vec{c}\,\Delta y\,\Delta z$. Assuming $\vec{F}$ is approximately constant on S_1, we have

$$\int_{S_1} \vec{F} \cdot d\vec{A} \approx \vec{F}(\vec{r}_0) \cdot \Delta\vec{A} = -\vec{F}(\vec{r}_0) \cdot (\vec{b} \times \vec{c})\,\Delta y\,\Delta z.$$

On S_2, the outward normal points in the other direction, so

$$\int_{S_1} \vec{F} \cdot d\vec{A} \approx \vec{F}(\vec{r}_0 + \vec{a}\,\Delta x) \cdot (\vec{b} \times \vec{c})\,\Delta y\,\Delta z.$$

Thus

$$\int_{S_1} \vec{F} \cdot d\vec{A} + \int_{S_2} \vec{F} \cdot d\vec{A} \approx (\vec{F}(\vec{r}_0 + \vec{a}\,\Delta x) \cdot (\vec{b} \times \vec{c}) - \vec{F}(\vec{r}_0) \cdot (\vec{b} \times \vec{c}))\,\Delta y\,\Delta z.$$

In other words, if f is the function $f(\vec{r}) = \vec{F}(\vec{r}) \cdot (\vec{b} \times \vec{c})$, then

$$\int_{S_1} \vec{F} \cdot d\vec{A} + \int_{S_2} \vec{F} \cdot d\vec{A} \approx (f(\vec{r}_0 + \Delta x \vec{a}) - f(\vec{r}_0)).$$

Now, by local linearity

$$f(\vec{r}_0 + \Delta x \vec{a}) \approx f(\vec{r}_0) + \operatorname{grad} f \cdot (\Delta x \vec{a}),$$

so

$$\int_{S_1} \vec{F} \cdot d\vec{A} + \int_{S_2} \vec{F} \cdot d\vec{A} \approx \operatorname{grad} f \cdot \vec{a}\,\Delta x\,\Delta y\,\Delta z = \operatorname{grad}(\vec{F} \cdot \vec{b} \times \vec{c}) \cdot \vec{a}\,\Delta x\,\Delta y\,\Delta z.$$

By an analogous argument, the contribution to the flux from S_3 and S_4 is approximately

$$\operatorname{grad}(\vec{F} \cdot \vec{c} \times \vec{a}) \cdot \vec{b}\,\Delta x\,\Delta y\,\Delta z$$

and the contribution to the flux from S_5 and S_6 is approximately

$$\operatorname{grad}(\vec{F} \cdot \vec{a} \times \vec{b}) \cdot \vec{c}\,\Delta x\,\Delta y\,\Delta z.$$

Thus, adding these contributions we have

Total flux through $S \approx (\operatorname{grad}(\vec{F} \cdot \vec{b} \times \vec{c}) \cdot \vec{a} + \operatorname{grad}(\vec{F} \cdot \vec{c} \times \vec{a}) \cdot \vec{b} + \operatorname{grad}(\vec{F} \cdot \vec{a} \times \vec{b}) \cdot \vec{c})\,\Delta x\,\Delta y\,\Delta z.$

The volume of the parallelepiped is $(\Delta x \vec{a}) \cdot ((\Delta y \vec{b}) \times (\Delta z \vec{c})) = \vec{a} \cdot (\vec{b} \times \vec{c})\,\Delta x\,\Delta y\,\Delta z$, so

$$\text{Total flux through } S \approx \operatorname{div} \vec{F}\,(\text{Volume of } S),$$

that is

$$(\operatorname{grad}(\vec{F} \cdot \vec{b} \times \vec{c}) \cdot \vec{a} + \operatorname{grad}(\vec{F} \cdot \vec{c} \times \vec{a}) \cdot \vec{b} + \operatorname{grad}(\vec{F} \cdot \vec{a} \times \vec{b}) \cdot \vec{c})\,\Delta x\,\Delta y\,\Delta z$$
$$\approx \operatorname{div} \vec{F}\,(\vec{a} \cdot (\vec{b} \times \vec{c}))\,\Delta x\,\Delta y\,\Delta z.$$

This is the formula which we are trying to prove holds, multiplied on both sides by $\Delta x\,\Delta y\,\Delta z$.

(b) The triple product $\vec{F} \cdot \vec{b} \times \vec{c}$ expands into 6 terms; therefore each component of its gradient has 6 terms, so $\operatorname{grad}(\vec{F} \cdot \vec{b} \times \vec{c})$ has 18 terms; there are three terms like this on the left-hand side, so the left-hand side has 54 terms.

(c) Since we are only interested in F_1, we only need to look at the first term in each of the dot products $\vec{F} \cdot \vec{b} \times \vec{c}$, $\vec{F} \cdot \vec{c} \times \vec{a}$, and $\vec{F} \cdot \vec{a} \times \vec{b}$. These terms are

$$F_1(b_2 c_3 - b_3 c_2), \quad F_1(c_2 a_3 - c_3 a_2), \quad F_1(a_2 b_3 - a_3 b_2).$$

Then, since we are only interested in the x partial derivative, we only need to look at the first term in each of the dot products $\operatorname{grad}(\cdots) \cdot \vec{a}$, $\operatorname{grad}(\cdots) \cdot \vec{b}$, and $\operatorname{grad}(\cdots) \cdot \vec{c}$. So we get 6 terms,

$$\frac{\partial F_1}{\partial x}((b_2 c_3 - b_3 c_2)a_1 + (c_2 a_3 - c_3 a_2)b_1 + (a_2 b_3 - a_3 b_2)c_1)$$
$$= \frac{\partial F_1}{\partial x}(b_2 c_3 a_1 - b_3 c_2 a_1 + c_2 a_3 b_1 - c_3 a_2 b a_1 + a_2 b_3 c_1 - a_2 b_3 c_1) = \frac{\partial F_1}{\partial x} \vec{a} \cdot \vec{b} \times \vec{c}.$$

(d) This starts out the same way, but this time we look at the second component in each gradient, so we get

$$\frac{\partial F_1}{\partial y}((b_2 c_3 - b_3 c_2)a_2 + (c_2 a_3 - c_3 a_2)b_2 + (a_2 b_3 - a_3 b_2)c_2)$$

$$= \frac{\partial F_1}{\partial y}(b_2 c_3 a_2 - b_3 c_2 a_2 + c_2 a_3 b_2 - c_3 a_2 b_2 + a_2 b_3 c_2 - a_2 b_3 c_2) = 0.$$

(e) The terms involving $\frac{\partial F_2}{\partial y}$ add up to $\frac{\partial F_2}{\partial y}(\vec{a} \cdot \vec{b} \times \vec{c})$ and those involving $\frac{\partial F_3}{\partial z}$ add up to $\frac{\partial F_3}{\partial z}(\vec{a} \cdot \vec{b} \times \vec{c})$. All the rest have some cancellation which makes them zero. So the whole thing works out to

$$\left(\frac{\partial F_1}{\partial x} + \frac{\partial F_2}{\partial y} + \frac{\partial F_3}{\partial z}\right) \vec{a} \cdot \vec{b} \times \vec{c} = \operatorname{div} \vec{F} (\vec{a} \cdot \vec{b} \times \vec{c}).$$

Solutions for Chapter 20 Review

1. Figure 20.5 shows a two dimensional cross-section of the vector field $\vec{v} = -2\vec{r}$. The vector field points radially inwards, so if we take S to be a sphere of radius R centered at the origin, oriented outward, we have

$$\vec{v} \cdot \Delta \vec{A} = -2R \|\Delta \vec{A}\|,$$

for a small area vector $\Delta \vec{A}$ on the sphere. Therefore,

$$\int_S \vec{v} \cdot d\vec{A} = \int_S -2R \|d\vec{A}\| = -2R(\text{Surface area of sphere}) = -2R(4\pi R^2) = -8\pi R^3.$$

Thus, we find that

$$\operatorname{div} \vec{v}(0,0,0) = \lim_{\text{vol} \to 0} \left(\frac{\int_S \vec{v} \cdot d\vec{A}}{\text{Volume of sphere}}\right) = \lim_{R \to 0} \left(\frac{-8\pi R^3}{\frac{4}{3}\pi R^3}\right) = -6.$$

Notice that the divergence is negative. This is what you would expect, since the vector field represents an inward flow at the origin.

Since $\vec{v} = -2\vec{r} = -2x\vec{i} - 2y\vec{j} - 2z\vec{k}$, the coordinate definition give

$$\operatorname{div} \vec{v} = \frac{\partial}{\partial x}(-2x) + \frac{\partial}{\partial y}(-2y) + \frac{\partial}{\partial z}(-2z) = -2 - 2 - 2 = -6.$$

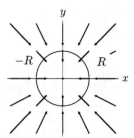

Figure 20.5: The vector field $\vec{v} = -2\vec{r}$

5. Since $\vec{r}$ is parallel to the slanted edges of the cone, the flux of $\vec{r}$ through the surface is all through the base (See Figure 20.6). On the base, $z = h$, and the normal is upward, so

$$\int_S \vec{r} \cdot d\vec{A} = \int_{\text{base}} (x\vec{i} + y\vec{j} + z\vec{k}) \cdot (dA\vec{k})$$
$$= \int_{\text{base}} h\, dA$$
$$= h(\text{Area of base})$$
$$= h(\pi b^2)$$

Thus

$$V = \frac{1}{3} \int_S \vec{r} \cdot d\vec{A} = \frac{\pi}{3} b^2 h.$$

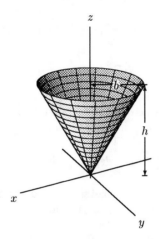

Figure 20.6

9. True. curl $\vec{F}$ is a vector whose value depends on the point at which it is calculated.

13. True. We calculate the x–components for each side of the equation:

$$(\text{curl}(f\vec{G}))_1 = \frac{\partial(fG_3)}{\partial y} - \frac{\partial(fG_2)}{\partial z}$$
$$= \frac{\partial f}{\partial y}G_3 + f\frac{\partial G_3}{\partial y} - \frac{\partial f}{\partial z}G_2 - f\frac{\partial G_2}{\partial z}$$
$$= \left(\frac{\partial f}{\partial y}G_3 - \frac{\partial f}{\partial z}G_2\right) + f\left(\frac{\partial G_3}{\partial y} - \frac{\partial G_2}{\partial z}\right)$$
$$= ((\text{grad } f) \times \vec{G})_1 + (f(\text{curl } \vec{G}))_1.$$

Computations for the other two components are similar, so

$$\text{curl}(f\vec{G}) = (\text{grad } f) \times \vec{G} + f \cdot (\text{curl } \vec{G}).$$

17. The flux of $\vec{E}$ through a small sphere of radius R around the point marked P is negative, because all the arrows are pointing into the sphere. The divergence at P is

$$\text{div } \vec{E} \, (P) = \lim_{\text{vol} \to 0} \left(\frac{\int_S \vec{E} \cdot d\vec{A}}{\text{Volume of sphere}} \right) = \lim_{R \to 0} \left(\frac{\text{Negative number}}{\frac{4}{3}\pi R^3} \right) \leq 0.$$

By a similar argument, the divergence at Q must be positive or zero.

21. Check that div $\vec{E} = 0$ by taking partial derivatives. For instance,

$$\frac{\partial E_1}{\partial x} = \frac{\partial}{\partial x}[q(x - x_0)[(x - x_0)^2 + (y - y_0)^2 + (z - z_0)^2]^{-3/2}]$$

$$= q[(y - y_0)^2 + (z - z_0)^2 - 2(x - x_0)^2][(x - x_0)^2 + (y - y_0)^2 + (z - z_0)^2]^{-5/2}$$

and similarly,

$$\frac{\partial E_2}{\partial y} = q[(x - x_0)^2 + (z - z_0)^2 - 2(y - y_0)^2][(x - x_0)^2 + (y - y_0)^2 + (z - z_0)^2]^{-5/2}$$

$$\frac{\partial E_3}{\partial z} = q[(x - x_0)^2 + (y - y_0)^2 - 2(z - z_0)^2][(x - x_0)^2 + (y - y_0)^2 + (z - z_0)^2]^{-5/2}.$$

Therefore,

$$\frac{\partial E_1}{\partial x} + \frac{\partial E_2}{\partial y} + \frac{\partial E_3}{\partial z} = 0.$$

The vector field $\vec{E}$ is defined everywhere but at the point with position vector $\vec{r}_0$. If this point lies outside the surface S, the Divergence Theorem can be applied to the region R enclosed by S, yielding:

$$\int_S \vec{E} \cdot d\vec{A} = \int_R \text{div } \vec{E} \, dV = 0.$$

If the charge q is located inside S, consider a small sphere S_a centered at q and contained in R. The Divergence Theorem for the region R' between the two spheres yields:

$$\int_S \vec{E} \cdot d\vec{A} + \int_{S_a} \vec{E} \cdot d\vec{A} = \int_{R'} \text{div } \vec{E} \, dV = 0.$$

In this formula, the Divergence Theorem requires S to be given the outward orientation, and S_a the inward orientation. To compute $\int_{S_a} \vec{E} \cdot d\vec{A}$, we use the fact that on the surface of the sphere, $\vec{E}$ and $\Delta \vec{A}$ are parallel and in opposite directions, so

$$\vec{E} \cdot \Delta \vec{A} = -\|\vec{E}\|\|\Delta \vec{A}\|$$

since on the surface of a sphere of radius a,

$$\|\vec{E}\| = q\frac{\|\vec{r} - \vec{r}_0\|}{\|\vec{r} - \vec{r}_0\|^3} = \frac{q}{a^2}.$$

Then,

$$\int_{S_a} \vec{E} \cdot d\vec{A} = \int -\frac{q}{a^2}\|d\vec{A}\| = \frac{-q}{a^2} \cdot \text{Surface area of sphere} = -\frac{q}{a^2} \cdot 4\pi a^2 = -4\pi q.$$

$$\int_{S_a} \vec{E} \cdot d\vec{A} = -4\pi q.$$

$$\int_S \vec{E} \cdot d\vec{A} - \int_{S_a} \vec{E} \cdot d\vec{A} = 4\pi q.$$

25. (a) Since $\vec{v} = \text{grad } \phi$ we have

$$\vec{v} = \left(1 + \frac{y^2 - x^2}{(x^2 + y^2)^2}\right)\vec{i} + \frac{-2xy}{(x^2 + y^2)^2}\vec{j}$$

(b) Differentiating the components of $\vec{v}$, we have

$$\text{div } \vec{v} = \frac{\partial}{\partial x}\left(1 + \frac{y^2 - x^2}{(x^2 + y^2)^2}\right) + \frac{\partial}{\partial y}\left(\frac{-2xy}{(x^2 + y^2)^2}\right) = \frac{2x(x^2 - 3y^2)}{(x^2 + y^2)^3} + \frac{2x(3y^2 - x^2)}{(x^2 + y^2)^3} = 0$$

(c) The vector $\vec{v}$ is tangent to the circle $x^2 + y^2 = 1$, if and only if the dot product of the field on the circle with any radius vector of that circle is zero. Let (x, y) be a point on the circle. We want to show: $\vec{v} \cdot \vec{r} = \vec{v}(x, y) \cdot (x\vec{i} + y\vec{j}) = 0$. We have:

$$\vec{v}(x, y) \cdot (x\vec{i} + y\vec{j}) = \left((1 + \frac{y^2 - x^2}{(x^2 + y^2)^2})\vec{i} + \frac{-2xy}{(x^2 + y^2)^2}\vec{j}\right) \cdot (x\vec{i} + y\vec{j})$$

$$= x + x\frac{y^2 - x^2}{(x^2 + y^2)^2} - \frac{2xy^2}{(x^2 + y^2)^2}$$

$$= \frac{x(x^2 + y^2 - 1)}{x^2 + y^2},$$

but we know that for any point on the circle, $x^2 + y^2 = 1$, thus we have $\vec{v} \cdot \vec{r} = 0$. Therefore, the velocity field is tangent to the circle. Consequently, there is no flow through the circle and any water on the outside of the circle must flow around it.

(d)

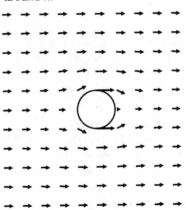

Figure 20.7

29. (a) Differentiating $\text{div } \vec{E} = 4\pi\rho$ with respect to time gives

$$\frac{\partial}{\partial t}\left(\text{div } \vec{E}\right) = \frac{\partial}{\partial t}\left(\frac{\partial E_1}{\partial x} + \frac{\partial E_2}{\partial y} + \frac{\partial E_3}{\partial z}\right) = 4\pi\frac{\partial\rho}{\partial t}.$$

Since, for example, $\frac{\partial}{\partial t}\left(\frac{\partial E_1}{\partial x}\right) = \frac{\partial}{\partial x}\left(\frac{\partial E_1}{\partial t}\right)$, we can rewrite this as

$$\frac{\partial}{\partial x}\left(\frac{\partial E_1}{\partial t}\right) + \frac{\partial}{\partial y}\left(\frac{\partial E_2}{\partial t}\right) + \frac{\partial}{\partial z}\left(\frac{\partial E_3}{\partial t}\right) = 4\pi\frac{\partial\rho}{\partial t}.$$

So we have shown that

$$\operatorname{div}\left(\frac{\partial \vec{E}}{\partial t}\right) = 4\pi \frac{\partial \rho}{\partial t}.$$

Now consider the equation

$$\operatorname{curl} \vec{B} - \frac{1}{c}\frac{\partial \vec{E}}{\partial t} = \frac{4\pi}{c}\vec{J}$$

and take the divergence of both sides:

$$\operatorname{div}\operatorname{curl} \vec{B} - \frac{1}{c}\operatorname{div}\left(\frac{\partial \vec{E}}{\partial t}\right) = \frac{4\pi}{c}\operatorname{div}\vec{J}.$$

Since $\operatorname{div}\operatorname{curl} \vec{B} = 0$, by Problem 16 on page 137, we have

$$-\frac{1}{c}\operatorname{div}\left(\frac{\partial \vec{E}}{\partial t}\right) = \frac{4\pi}{c}\operatorname{div}\vec{J}.$$

Thus

$$-\frac{1}{c}\left(4\pi\frac{\partial \rho}{\partial t}\right) = \frac{4\pi}{c}\operatorname{div}\vec{J},$$

so

$$-\frac{\partial \rho}{\partial t} = \operatorname{div}\vec{J},$$

or

$$\frac{\partial \rho}{\partial t} + \operatorname{div}\vec{J} = 0.$$

(b) The equation derived in part (a) says that the rate of change with time of the charge density at a point is the negative of the divergence of the current density at that point.

Why is this reasonable? Suppose $\operatorname{div}\vec{J} < 0$ at some point, that is, there is a current "sink" there. This means that current is "piling up" at this point – in other words, the charge is "piling up" there. Thus we would expect $\partial\rho/\partial t > 0$ there. Similarly, if $\operatorname{div}\vec{J} > 0$, at some point, there is a current source there. This means that current is being "created" near that point, which means that charge density is decreasing there. Thus we would expect $\partial\rho/\partial t < 0$ there.

(c) The equation is called the charge conservation equation because it reflects the fact that charge is neither created nor destroyed. If $\operatorname{div}\vec{J}$ is negative at some point, there is a net influx of current into a small surface around the point, so the charge density must be increasing there. If $\operatorname{div}\vec{J}$ is positive, there is a net outflow of current through a small surface around the point, so the charge density must be decreasing there.

APPENDIX

Solutions for Section D

1.

$$\int (x^2 + 2x + \frac{1}{x}) \, dx = \int x^2 \, dx + \int 2x \, dx + \int \frac{1}{x} \, dx$$
$$= \frac{1}{3}x^3 + x^2 + \ln|x| + C,$$

where C is a constant.

5. Let $2t = w$, then $2dt = dw$, so $dt = \frac{1}{2}dw$, so

$$\int \cos 2t \, dt = \int \frac{1}{2}\cos w \, dw = \frac{1}{2}\sin w + C = \frac{1}{2}\sin 2t + C,$$

where C is a constant.

9. Let $t^2 + 1 = w$, then $2tdt = dw$, $tdt = \frac{1}{2}dw$, so

$$\int te^{t^2+1} \, dt = \int e^w \cdot \frac{1}{2} \, dw = \frac{1}{2}\int e^w \, dw = \frac{1}{2}e^w + C = \frac{1}{2}e^{t^2+1} + C,$$

where C is a constant.

13. Let $\cos 5\theta = w$, then $-5\sin 5\theta d\theta = dw$, $\sin 5\theta d\theta = -\frac{1}{5}dw$. So

$$\int \sin 5\theta \cos^3 5\theta \, d\theta = \int w^3 \cdot (-\frac{1}{5}) \, dw = -\frac{1}{5}\int w^3 \, dw = -\frac{1}{20}w^4 + C$$
$$= -\frac{1}{20}\cos^4 5\theta + C,$$

where C is a constant.

17.

$$\int xe^x \, dx = xe^x - \int e^x \, dx \qquad (\text{let } x = u, e^x = v', e^x = v)$$
$$= xe^x - e^x + C,$$

where C is a constant.

21. Let $x^2 = w$, then $2xdx = dw$, $x = 1 \Rightarrow w = 1$, $x = 3 \Rightarrow w = 9$. Thus,

$$\int_1^3 x(x^2 + 1)^{70} \, dx = \int_1^9 (w + 1)^{70}\frac{1}{2} \, dw$$
$$= \frac{1}{2} \cdot \frac{1}{71}(w + 1)^{71}\Big|_1^9$$
$$= \frac{1}{142}(10^{71} - 2^{71}).$$

25. Let $\sqrt{x} = w$, $\frac{1}{2}x^{-\frac{1}{2}}\,dx = dw$, $\frac{dx}{\sqrt{x}} = 2\,dw$. Then $x = 1 \Rightarrow w = 1$, $x = 4 \Rightarrow w = 2$. So

$$\int_1^4 \frac{e^{\sqrt{x}}}{\sqrt{x}}\,dx = \int_1^2 e^w \cdot 2\,dw = 2e^w \Big|_1^2 = 2(e^2 - e) \approx 9.34.$$

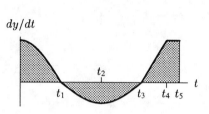

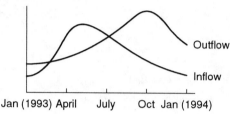

Figure D.1   Figure D.2

29. (a) Suppose we choose an x, $0 \le x \le 2$. If Δx is a small fraction of a meter, then the density of the rod is approximately $\rho(x)$ anywhere from x to $x + \Delta x$ meters from the left end of the rod.

    The mass of the rod from x to $x + \Delta x$ meters is therefore approximately $\rho(x)\Delta x$.

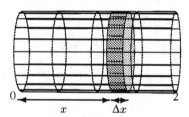

(b) The definite integral is

$$M = \int_0^2 \rho(x)\,dx = \int_0^2 (2 + 6x)\,dx = (2x + 3x^2)\Big|_0^2 = 16 \text{ grams.}$$

33. (a) Partition $0 \le h \le 100$ into N subintervals of width $\Delta h = \frac{100}{N}$. The density is taken to be approximately $\rho(h_i)$ on the ith spherical shell, and the volume is approximately the surface area of a sphere of radius $r_e + h_i$ meters times Δh, where $r_e = 6.37 \times 10^6$ meters is the radius of the earth. If the volume of the ith shell is V_i, then $V_i \approx 4\pi(r_e + h_i)^2\Delta h$, and a left-hand Riemann sum for the total mass is

$$M \approx \sum_{i=0}^{N-1} 4\pi(r_e + h_i)^2 \times 1.28e^{-0.000124h_i}\Delta h.$$

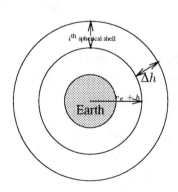

Figure D.3

(b)   This Riemann sum becomes the integral

$$M = 4\pi \int_0^{100} (r_e + h)^2 \times 1.28e^{-0.000124h} \, dh$$

$$= 4\pi \int_0^{100} (6.37 \times 10^6 + h)^2 \times 1.28e^{-0.000124h} \, dh.$$

Evaluating the integral using numerical methods gives $M = 6.48 \times 10^{16}$ kg.

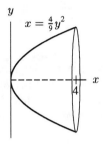

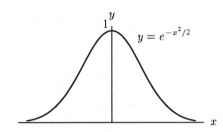

Figure D.4   **Figure D.5**

37.   We want to approximate $\int_0^{120} A(h) \, dh$, where h is height, and $A(h)$ represents the cross-sectional area of the trunk at height h. Since $A = \pi r^2$ (circular cross-sections), and $c = 2\pi r$, where c is the circumference, we have $A = \pi r^2 = \pi[c/(2\pi)]^2 = c^2/(4\pi)$. We make a table of $A(h)$ based on this:

TABLE D.1

height (feet)	0	20	40	60	80	100	120
Area (square feet)	53.79	38.52	28.73	15.60	2.865	0.716	0.080

We now form left & right sums using the chart:

$$\text{LEFT}(6) = 53.79 \cdot 20 + 38.52 \cdot 20 + 28.73 \cdot 20 + 15.60 \cdot 20 + 2.865 \cdot 20 + 0.716 \cdot 20$$
$$= 2804.42.$$
$$\text{RIGHT}(6) = 38.52 \cdot 20 + 28.73 \cdot 20 + 15.60 \cdot 20 + 2.865 \cdot 20 + 0.716 \cdot 20 + 0.080 \cdot 20$$
$$= 1730.22$$

So

$$\text{TRAP}(6) = \frac{\text{RIGHT}(6) + \text{LEFT}(6)}{2} = \frac{2804.42 + 1730.22}{2} = 2267.32 \text{ cubic feet.}$$

Solutions for Section F

1.

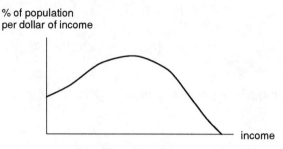

Figure F.6: Density function

5. (a) Most of the earth's surface is below sea level. Much of the earth's surface is either around 3 miles below sea level or exactly at sea level. It appears that essentially all of the surface is between 4 miles below sea level and 2 miles above sea level. Very little of the surface is around 1 mile below sea level.

   (b) The fraction below sea level corresponds to the area under the curve from -4 to 0 divided by the total area under the curve. This appears to be about $\frac{3}{4}$.

9. (a) i.   ii.

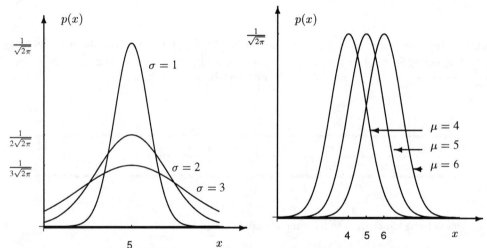

   (b) Recall that the mean is the "balancing point." In other words, if the area under the curve was made of cardboard, we'd expect it to balance at the mean. All of the graphs are symmetric across the line $x = \mu$, so μ is the "balancing point" and hence the mean.

   As the graphs also show, increasing σ flattens out the graph, in effect lessening the concentration of the data near the mean. Thus, the smaller the σ value, the more data is clustered around the mean.

Solutions for Section G

1. $(1,0)$

5. $\left(\frac{5\sqrt{3}}{2}, -\frac{5}{2}\right)$

9. $r = \sqrt{0^2 + 2^2} = 2, \quad \theta = \pi/2.$

13. $r = \sqrt{(0.2)^2 + (-0.2)^2} = 0.28.$

$\tan \theta = 0.2/(-0.2) = -1.$ Since the point is in the fourth quadrant, $\theta = 7\pi/4.$ (Alternatively $\theta = -\pi/4.$)

Notes

Notes

Notes

Notes

Notes

Notes

Notes

Notes

Notes

Notes

Notes